To
Kathy

Enjoy the read.

Wayne Smith
15/12/2007.

MW01688153

Ripper Tassie Place Names

P&WS
PUBLISHING

P&WS Publishing
wgsmith@iprimus.com.au

ISBN 978-0-9804419-0-1

Printed by Foot & Playsted Fine Printers, Launceston, info@footandplaysted.com.au

Design and typography by Dean Smith, deanism@internode.on.net
Set in Nueva 11/13

Ripper Tassie Place Names

"The more you read, the more ignorant you become"

John Saunders
(Stepfather of the author).

Contents

Definitions of Tassie Terms

Before launching seriously into Tasmania's wonderful array of ripper place names, it should be understood that many terms used for geographical features may be peculiar to Tasmania. Brief explanations for these terms are listed below:

BOTTOM	=	low lying land.
GUT	=	narrow passage between sea cliffs.
GUTTER	=	ditch, trough or channel for the carriage of water.
HOLE	=	hollow, pit, cave, fishing spot, hunting spot.
HOLLOW	=	cavity, hole, depression, valley.
POT	=	naturally occurring hole, pit, depression etc. all generally containing water.
QUOIN	=	a wedge-shaped peak.
SUGARLOAF	=	a cone-shaped hill or peak reminiscent of a cone of sugar.
SWALLET	=	an abyss (usually a cave) into which a stream disappears.
TIERS	=	long flat-topped range of mountains with sheer cliff faces (as in a tier of a cake).
TOR	=	a rocky eminence.

Sources

Unless stated otherwise, all of the definitions used in this book are from Websters Dictionary

All of the information on British etymology has come from three basic sources:

1. *Oxford Dictionary of English Place Names* (AD Mills)
2. *Dictionary of Place Names in the British Isles* (Adrian Room)
3. *Place Names of Great Britain and Ireland* (John Field)

Tasmanian Aboriginal place names were primarily sourced from *Tasmanian Place Names – the Aboriginal Connection* (John A Taylor)

Abbreviations

AOT	Archives of Tasmania
Br	Brittonic (Brythonic)
c	Circa
C	Century
Capt.	Captain
CE	Common, Current or Christian Era
E	East
Fr	French
Gael	Gaelic
HEC	Hydro Electric Commission
HTG	Hobart Town Gazette
HWC	Hobart Walking Club
Ir	Irish
L	Latin
Lt.	Lieutenant
MHA	Member of the House of Assembly
MLC	Member of the Legislative Council
N	North
OE	Old English
OF	Old French
OS	Old Scandinavian
S	South
UK	United Kingdom
W	West
WW1	World War 1
WW2	World War 2

See Location Guide next page for location abbreviations.

Location Guide

BS	**Bass Strait**	King and Flinders Islands groups
CH	**Central Highlands**	Bounded by Bothwell, Liffey, Cradle Mountain and Interlaken
EC	**East Coast**	Bicheno to Orford
NC	**North Coast**	Bridport to Port Sorell
NE	**North East**	Bridport to Bicheno
NM	**Northern Midlands**	Campbell Town area
Nt	**Northern Tasmania**	Launceston, Tamar Valley and surrounds
NW	**North West**	Port Sorell to Arthur River
SC	**South Coast**	Coastal features South Cape to South West Cape
SE	**South East**	Orford to Cockle Creek
SM	**Southern Midlands**	Oatlands area
St	**Southern Tasmania**	South of Gordon River Road
SW	**South West**	South West National Park and Hartz Mountains
WC	**West Coast**	Arthur River to Macquarie Harbour

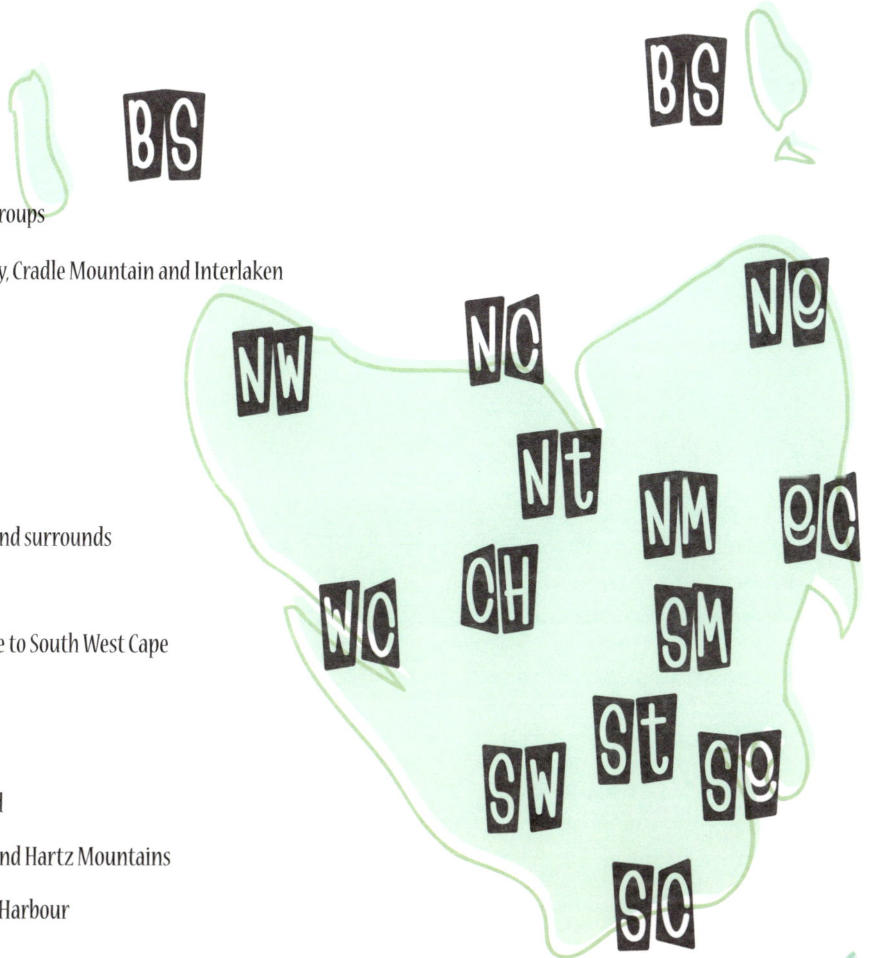

Acknowledgements

I am heavily indebted to the illustrator Colin Abel for his brilliant and highly amusing cartoons and to Dean Smith for his superb graphic art and photographic images, to photographers Craig Hoehne and Peter Daalder for some of the photos in this production and to my good friends John Haughton, Bruce Andrew and Robin Barker for editing the text.

I am is also indebted to the Nomenclature Board (in particular Tony Naughton and Mick Casey), for their assistance regarding obscure names in this book and for giving me access to nomenclature records over many years to ascertain the 'official' origins and locations of numerous place names. I have also gathered additional information from the Archives Office of Tasmania and the Tasmaniana Library. The staff members of both of these organisations have been incredibly helpful in assisting me to trace 'hidden' records on nomenclature.

I have appeared regularly on ABC Radio's Nomenclature Talk-back on some Saturday mornings for around five years and this has been a fruitful source of information and history behind Tasmanian place names. I thank many listeners for freely providing origins of place names and I am grateful to the ABC for allowing me to appear on radio programs to give me access to the general public for such valuable information. In particular, I extend special thanks to ABC presenter Chris Wisbey who 'discovered' me several years ago.

Finally, I appreciate the love and support given by my wife Pauline for allowing me the freedom to spend endless hours compiling this information.

Wayne Smith.

Origins of place names recorded on the front cover

Nowhere Else	– **See pages 10-11**	**Bust Me Gall Hill**	– **See page 64**
Dubbil Barril	– **See page 58**	**Lake Leake**	– **See page 73**
Whatucallit Road	– **See page 8**	**Lovely Bottom**	– **See page 80**

John Albert Taylor
(17/9/1933 – 26/8/2007)

This book is dedicated to the late John Albert Taylor who died 26th August, 2007. John and the author have spent a collective 50 years of research into the origins of place names. They jointly produced an unpublished book called *Tasmanian Place Names* (1993) which can be accessed in the Tasmaniana Library, principally for those people wanting information for family history research. In recent years John has become the foremost linguist in the Tasmanian Aboriginal (Palawa) languages and was recently rewarded with a Masters degree at the University of Tasmania for his work in this area. Some of the information in this book (particularly with regard to the origins of Aboriginal place names) has been sourced from John's research.

Introduction

Warning!

This book is filled with awful puns and gratuitous, politically incorrect comments in an attempt to have fun with some of Tasmania's many unique names. Most books on place names are presented in a lacklustre, dictionary format, and the author has taken considerable pains to avoid this pitfall. Consequently, in an attempt to bring history alive, this publication introduces quirky humour into the text to make the book more interesting and amusing to read. Nevertheless, the book has a serious side as well, because I have endeavoured to provide reliable information on the origins of most of the unusual names recorded in this publication. Where the etymology of a name is not known, 'educated guesses' were made based on years of experience, so that each name presented, has been given a logical reason for its origin. It is hoped that the levity of the notes and illustrations will not be seen as frivolous, as I take place names very seriously and do not seek to trivialize the importance of our wonderful nomenclature heritage.

Hidden Names

The most amusing and unusual names in Tasmania are 'hidden', as they are not shown on maps or road signs. Some examples are: names of caves, old disused mines and minor creeks, swamps, gullies, hills and lakes. Likewise, the original Aboriginal name for many areas is either not generally publicised, or is not used. Consequently, the majority of us are oblivious to this wonderful untapped heritage in nomenclature. Therefore, the primary reason that this book was conceived was to inform residents and visitors alike, about Tasmania's largely secret heritage in weird, wonderful and wacky names.

Nomenclature Pitfalls

Ascertaining the origins of place names is not an exact science. Ever since European settlement, names have appeared on maps, usually without explanations. Therefore, in many cases, unearthing the origin of a name can prove difficult, if not impossible. Consequently, the origins of the names given in this book are frequently arrived at by a process of elimination.

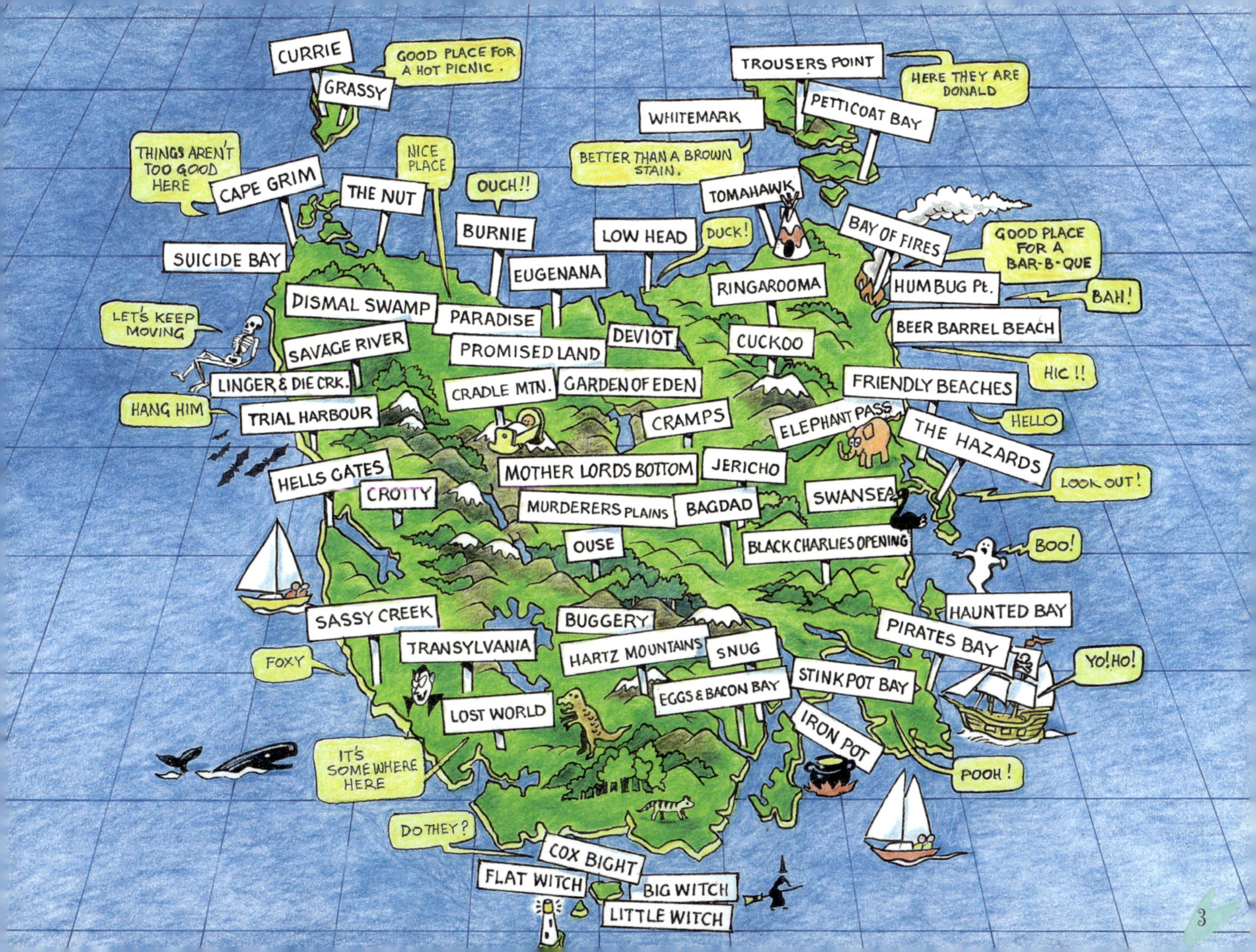
CURRIE
GOOD PLACE FOR A HOT PICNIC.
GRASSY
TROUSERS POINT
HERE THEY ARE DONALD
PETTICOAT BAY
WHITEMARK
BETTER THAN A BROWN STAIN.
THINGS AREN'T TOO GOOD HERE
CAPE GRIM
NICE PLACE
THE NUT
OUCH!!
TOMAHAWK
BURNIE
LOW HEAD
DUCK!
BAY OF FIRES
GOOD PLACE FOR A BAR-B-QUE
SUICIDE BAY
EUGENANA
RINGAROOMA
HUMBUG Pt.
BAH!
DISMAL SWAMP
PARADISE
LET'S KEEP MOVING
DEVIOT
CUCKOO
BEER BARREL BEACH
SAVAGE RIVER
PROMISED LAND
HIC!!
LINGER & DIE CRK.
CRADLE MTN.
GARDEN OF EDEN
FRIENDLY BEACHES
HANG HIM
TRIAL HARBOUR
CRAMPS
ELEPHANT PASS
HELLO
THE HAZARDS
MOTHER LORDS BOTTOM
JERICHO
HELLS GATES
LOOK OUT!
CROTTY
SWANSEA
MURDERERS PLAINS
BAGDAD
OUSE
BLACK CHARLIES OPENING
BOO!
HAUNTED BAY
SASSY CREEK
BUGGERY
PIRATES BAY
TRANSYLVANIA
HARTZ MOUNTAINS
SNUG
FOXY
YO! HO!
STINKPOT BAY
EGGS & BACON BAY
LOST WORLD
IRON POT
IT'S SOME WHERE HERE
POOH!
DO THEY?
COX BIGHT
FLAT WITCH
BIG WITCH
LITTLE WITCH

Having Fun With Names

It is quite easy to poke fun at some of Tasmania's wonderful place names. Visitors shouldn't be fooled by Tasmania's evocative place names as you can experience hell at **PARADISE** and encounter paradise at **HELLS GATES**. You can ring for a room at **RINGAROOMA**, but you can't hire a camel at **DROMEDARY**; you might make a buck at **BUCKLAND** but you will probably get diddled at **DIDDLEUM PLAINS**. **WAPPING** is not whopping at all, in fact it is quite small, and paradoxically, so is **PLENTY**. In addition, there's not much doing at **DOO TOWN** (unless you like fishing), there is no bungy jumping at **ADVENTURE BAY** and things are not necessarily pear-shaped at **PEARSHAPE**. But please note that **SAVAGE RIVER** doesn't 'roar' (unless it is in flood), **CUCKOO** doesn't 'cuckoo' and **BELL BAY** doesn't 'ring'. On the other hand, be prepared to be shocked that **THE GRUNTER** (a hill) actually 'grunts' (see page 32), **DOG BARK** (a district) can 'bark' (as retreating surf from a gulch makes a barking sound), and **SQUEAKING POINT** really does 'squeak' (see page 32). I am even prepared to believe that **SQUEALING CREEK** may 'squeal' (but only in fright when it falls over a precipice). Despite the above, you can be assured that **LAKE LEAKE** doesn't leak, and (unless it rains) there is no ooze at **OUSE** and we know that **COOEE** is within a cooee of **BURNIE**. However, you should be aware that it isn't polite to burp at **BELCHERS LOOKOUT** and that **SNUG** is not always snug. In addition, you probably won't get cramps at **CRAMPS**, and you won't find work at **TRYWORK POINT** (because it is a long-disused whaling station). There are no needles in the haystacks at **NEEDLES** (even if you **can** find a haystack) and if your kids are missing, try **UPPER TEA TREE**, there is a good chance you'll find them up there. You also might be disappointed to discover that **BEAUTY POINT** was named after a dead cow called 'Beauty' (who was buried there - and is now a 'Sleeping Beauty'). The scenery there is quite pretty, but the 'real' beauty was bovine - no bull. Be aware that **NIERINNA** is pronounced 'Ny-er-inna' (not 'near in 'er'), **RAMINEA** is pronounced 'Ram-in-ee-ah' (not 'Ram in ear') and naturally, **MIKE HOWES NECK** should not sound like 'My Cow's Neck'. Most confusing of all, **TARGA**, Tasmania is **not** Targa Tasmania. The former is a town named after a local Aboriginal word 'to cry' whilst the latter (Targa Tasmania touring car rally) has an Italian origin and has no connection with **TARGA**, Tasmania. So the moral is, unless you are racing in Targa Tasmania, don't try to speed through **TARGA**, Tasmania. Finally, there is no good wood at **GOODWOOD** which begs the question:

'How much good wood – would a wood cutter cut – if a wood cutter would – cut good wood good?'

Despite all of the above, if you are near Cradle Mountain and you are planning boating activities, please be aware that you can't go up **SCHIST CREEK** without a paddle.

As you can see, you can have heaps of fun with Tasmanian place names, particularly if you have a devious mind like mine.

Bogaduck Speedway Whitemark, Flinders Island BS

This fully-licensed racing circuit was built for stock car racing by members of the local Automobile Club *'on land so boggy that it would bog a duck'*. By all reports, following wet weather, the racing here is a sight to behold.[1] When I saw Colin Abel's brilliant cartoon of Bogaduck, I couldn't resist penning the following quip:

'Boga' Duck, with lots of pluck, and a tick of luck, could make a quick buck, with his towing truck, plucking stuck-up ducks, with quacks and clucks, from the Bogaduck muck.

NB. Shucks, this poem sucks.

1 *Flinders Island* (Jean Edgecombe) p 47

Tas Mania

Who was the idiot who coined the name Tasmania? What right-thinking person would inflict on us a place name that terminates in 'mania'? This implies that the inhabitants of our lovely island are Tas-maniacs. How embarrassing is that? The island's original name of Van Diemen's Land was applied by Abel Tasman in 1642 to honour his boss, the Governor of the Dutch East India Company. Because of the negative connotations of convict transportation, the island's name was changed to Tasmania on the 1st January 1856 to honour its 'discoverer'. However, Tasmanians would be even more horrified if they realised that the surname of Tasman (pronounced Tuss-man in Dutch) is an occupational name meaning 'bag-man', derived from the Old French word tasse 'a bag for holding things' i.e. a purse.[2] Therefore, if we anglicised the name of Tasmania, our island would then be called 'Bagmania' and Tasmanians (male and female) would then be branded as '[French] handbag-wielding maniacs'. The nickname of the state would then become 'Baggy' instead of 'Tassie'. The alternative of course, would be to reintroduce Tasman's appellation of Van Diemen's Land and we could then all be called 'Van Demons' (which is not much better). I could think of nothing worse than Tasmanians being depicted as a bunch of crazy crusty-demons, speeding around in vans or caravans. Tasmania would then risk being portrayed as one big smash-up derby. Instead of 'Explore the possibilities', our motto could then be 'Get smashed in Van Diemen's Land'.

2 *The New Holland Dictionary of Names* (William TS Noble)

PURSEY
BAGGY
MANIAC

Only In Tasmania...

Nowhere else in the world, will you find so many wacky, weird and wonderful place names as you will find in the island state of Tasmania. Not only is Tasmania fortunate to have so many ripper place names, it is doubly blessed with having dozens of 'nothing' names like Nowhere Else. The following names are just a few examples:

Lake Nameless Western Tiers NW

Answer this if you can. If this lake has a name (albeit 'Nameless') – how can it be nameless? Lake Nameless was named by Chambers and others in the early 1900s, in the belief that the lake had not been named. However, they did not realise that it had previously been named Lake Babylon and is sometimes still referred to by this name. So we still babble-on about Babylon whilst the lake is really nameless.[3] There was also a Mount Nameless in the Mt. Anne region SW, but this has since been renamed Eve Peak, after Eve Masterman, the first woman to scale this peak when she was a member of the Hobart Walking Club. So Mt. Nameless is no longer nameless. I quietly ponder if there are any serpents on Eve Peak.

Whatucallit Road

Florentine Valley, Upper Gordon River SW

Believe it or not, this is an official name, probably given by Whatshisname. The name was applied by Forestry Commission workers in the 1980s to part of an unfinished convict-built road (supervised by William Dawson) which ran from Hamilton to Gordon Bend. The Government-funded road was built in the early 1850s through the South West wilderness to open up Port Davey for settlement. Due to lack of funds during a depression, further construction of the road was abandoned and it terminated abruptly in the middle of the South-West wilderness near the big bend in the Gordon River. The road was never officially named but was commonly referred to as Dawsons Road. It was subsequently dubbed derisively 'The road to nowhere'. When a section of the road was rebuilt by Forestry workers 130 years later to access stands of timber, the original name of the road (if any) had been long forgotten, so they 'invented' the name 'Whatucallit' to fill the void.[4]

The Never Never Lookout

near Cradle Mountain CH

You'll never, never go there, if you never, never know where (it is); and you'll never, never see anything, if you come here when it is misty. This lookout gives awesome views over the wilderness (on a clear day). The name 'Never Never' is an apt one for such a wild desolate alpine area.

3 Nomenclature Board records

4 Advice from Bruce Reine (Forestry Tasmania) 2nd April 2007

Nonesuch Road Wattle Hill, Sorell SE

Nonsense, there is such a road, because I have driven along it. The road once accessed 'Nonsuch' (note spelling), a property located on land originally granted to pioneer Thomas Lascelles in 1813 and later acquired by Silas Gatehouse as part of his large 'Wattle Hill' estate. Early maps and road signs record the property name incorrectly as 'Nonesuch' and the erroneous spelling survived.

None Such Creek, (which does exist) flows through the 'Nonsuch' property.[5]

NB. The property was named after the Nonsuch Palace built 1538-45 by King Henry VIII as a hunting lodge in Surrey, at Cuddington, near Epsom, England. The cost of building and furnishing Nonsuch Palace was £24,536 which was literally a 'king's ransom' in those days with the result that the palace was the best in the land. The word 'nonsuch' means the opposite to what you might think as it stands for 'without equal, unparalleled, unrivalled, unsurpassed'.[6]

5 *Municipality of Sorell Heritage Study* Volume 6

6 http://www.epsom-ewell.gov.uk/Nonsuch/

Nowhere Else South of Barrington NW

This zany name was coined in the 1920s because the original road to the area came to a dead end at Charles Ivory's 'Nowhere Else' farm. The area around the farm eventually adopted the name of the farm which was obviously named as the road ended there and travellers could go nowhere else. The present link road now goes through Nowhere Else, so it is no longer a dead end route.

The road sign here is Tasmania's most photographed sign, so much so, that souvenir hunters keep stealing it and it has to be regularly replaced. A plaque telling the story of the name's origin was also stolen.[7] There is another location called No Where Else near Fingal Tier NE where the original road came to a dead end at another 'Nowhere Else' farm.

7 Source – Ray Chilcott (former owner of the farm) and Dave Lockwood (former neighbour to the farm)

NO WHERE ELSE

Zero Davey Street Hobart CBD

Can you believe that Hobart has a building listed as 'Zero'? This implies that the building is on a block of land that does not exist. This is the only known example of any city address in the world being recorded as zero (apart from Ground-Zero, New York). This catchy term is used for a modern and controversial block of buildings at the eastern entrance to Hobart's CBD which is completely out of character with the other 'old world' Edwardian buildings in the area (see the opposite page).

Apart from using a ridiculous term like 'Zero', the address is wrong anyway. The first building in Davey Street is an impressive Edwardian sandstone building known as the *Royal Engineers Building* which is listed as 2 Davey Street (see photos below and opposite). There are at least three other buildings between 2 Davey Street and the building titled 'Zero'. Consequently, Zero Davey Street buildings should be a higher number than 4 (possibly 10 or 12).

Photos Dean Smith

Are you confused?

Good, this means that I have gained your attention!

Please read on as there are some more confusing and intriguing 'nothing' names on the following pages.

ROYAL ENGINEERS BUILDING
2 DAVEY STREET

blu edge

All or Nutting

Nutting is the Irish pronunciation of 'nothing', so here are some nutting names. Nutting ventured, nutting gained.

Lake Nutting **Walls of Jerusalem N.P.** NW

The lake was named after Bill Nutting, a stockman and snarer who died here in 1927.

Nuttings Road **Meander** Nt

Is also named after William Nutting an early settler who is probably the same person.[8]

No Tea Creek **Shannon River** CH

Obviously the discoverer of the creek missed out on a meal. This used to happen to me when I was a boy, particularly when I had been rather naughty!

Notuck Hill **Lake Sorell** CH

Somebody else must have missed out on tucker, although I suppose it could indicate that Friar Tuck does not live around here.

No Mans Creek

Temma, Sandy Cape NW

This is probably no woman's creek as well. The creek was seemingly named to indicate that nobody had a mining lease on this creek. In other words it was open territory.

Nix Creek **Waratah** NW

This creek amounts to nothing. At least that is what some prospector thought when he worked the creek and found no minerals here.

No Duck Lagoon

Lady Barron, Flinders Is. BS

This sounds like a batsman's paradise. Actually, the negative name probably infers that duck hunters would be better served to try other lagoons.

No Good Bay **Maria Is.** EC

Why is this bay 'no good'? Well, it is mission impossible to access it for the mooring of boats.

8 *Tasmanian Place Names* (Taylor-Smith)

NUTTING
NUTTING
+ NO TEA
+ NO TUCK
+ NO MAN
+ NIX
+ NO DUCKS
+ NO GOOD
= NUTTING
IF YOU HAVE NUTTING TO SAY, SAY NUTTING.

Half Names

Something just as bad as 'nothing names' is 'half names' – as half of 'nothing' = 'nothing'. For example how would Napoleon have felt if his name had been shortened to 'Nappy', Nelson wouldn't have liked being called 'Nelly' and I daresay Churchill wouldn't appreciate 'Churchy'. But things are different here in Tasmania – we have quite a few 'half-names' (see the list below).

Half Lemon Rock

Freycinet Peninsula EC

A lemon is something that is no good (e.g. 'my car is a lemon'). So how good is half a lemon? Half a lemon is still a lemon. Seriously, this rock resembles the shape of half a lemon and it is known to fishermen as Broken (Lemon) Rock. There is another rock nearby known as Lemon Rock (which is 'the full lemon').

Half Rung Sugarloaf

Spring Bay EC

If someone half rings you, you only get half a message or no message at all. The message with this name, is that it denotes a half rung (ring) marking on this conical hill.

Half Woody Hill **Cox Bight** SC

Recently I saw a baseball hat in a service station shop with the message on it 'Crack a Woody'. I had never heard the term before and I asked the female attendant what it meant. She fell about laughing and eventually she sheepishly told me that it meant a male erection. I thought at first she was talking about a shed but now that I am wise in these matters, I pose the question – how much use is a 'Half-Woody'? In this case the answer is – 'quite useful' as the hill was named in 1956 because it was partially clad with trees. [9]

Halfpenny Hill **Arthurs Lake** CH

This is a clever name as this hill is a small spur on Mt. Penny (which is a good place to play *Two-up*). In other words – Halfpenny Hill is not the full penny. For the record, Mt Penny gained its name when government surveyor John Helder Wedge (what a great name) placed a penny on a nearby peak. Maybe he went there to 'spend a penny'. [10]

9 Nomenclature Board records
10 Nomenclature Board records

Arfa Circus Battery Point, Hobart SO

'Arfa Circus' is the local pronunciation of Arthur Circus. This sounds silly. Imagine going to half a circus! You would be only half-satisfied as there would be more clowns in the audience than in the ring (or should that be half-ring?).

Seriously, in England, a circular street is frequently known as a 'circus' e.g. Piccadilly Circus. In this case Runnymede Street runs into a circular village green surrounded by houses. Arthur Circus was named in memory of ex-governor Sir George Arthur (who once owned the land).

Awesome Stacks

Cape Hauy (pronounced Hoy) Fortescue Bay SE

Hoy! What sort of name is this? This cape was named in 1802 by Lt. Peron of the Baudin expedition after Rene-Just Hauy, French mineralogist and founder of the science of crystallography. Cape Hauy is renowned for its astonishing collection of jaw-dropping dolerite sea stacks at the tip. The spectacular cliffs here are over 200 metres high. Some of the more prominent rock features on the cape are:

Photo Dean Smith

The Candlestick

This would be a good place to hold 'Carols by Candlelight'. When you visit this place you should bring an old flame with you because the candle had gone out when I last visited. As the name suggests, this is a massive, roughly cylindrical, spectacular dolerite tower that rises up vertically 200 metres above the water. No other rock tower can hold a candle to this one for rock climbing as you would probably need candle wax to get a firm grip. It's wick-ed.

Totem Pole

This is a squarish, rock tower rising sheer to 65 metres that is a favourite for rock climbers because of the smooth, vertical rock faces. Red Indians say 'How?', but rock climbers say 'Wow!'

Cape Hauy   Photo Dean Smith

Totem Pole (left) and the Candlestick (center)

Photo Dean Smith

The Lanterns

I expected to find a lighthouse here, but the lights were out and there was nobody home (except for a few seagulls). The Lanterns are a series of mammoth rock towers rising sheer from the water to a height exceeding 200 metres. The Lanterns were obviously named at dawn or sunset when the vertical cliff faces were 'lit up' by the rays of the sun.

The Moai North of Fortescue Bay

This slender cenotaph-like sea stack was named after one of monolithic human figures known as Moais that were carved from rock on Easter Island, mostly between 1250 and 1500 CE. This stack is also irresistible for rock climbers.[11]

Above: The Moai with rock climbers on the tip

Right: The Moai with the Lanterns of Cape Hauy in the background.

Photos Mark Anderson (US rock climber)

11 http://en.wikipedia.org/wiki/Moai

Hippolyte Rocks (local pronunciation = Hippo-light; Greek = Hip-pol-it-tee) off Cape Hauy SE

Hippolyte? Who's kidding who? I've never yet seen a light hippo so the person who named these rocks must have been heavy-handed. The main island in this group of rocks is affectionately known as 'The Hippo' as it is an abbreviation of Hippolyte and because its shape vaguely resembles the back of an immersed hippopotamus. However, the name of the group of rocks has nothing to do with hippos. They were named by French navigators of the Bruny d'Entrecasteaux expedition 1793 presumably in honour of Hippolyte Deslacs (1777-1805), an apprentice aboard the *Recherche*. Cape Deslacs south of Cremorne was also named in honour of him. His full name has hippo-like dimensions as it is Charles-Francois Hippolyte Deslacs d'Arcambal. Hippolyte was the godson of Fleurieu the French Minister of Marine who sponsored the expedition.[12] Hippolyte derives his strange name from Hippolyte, the queen of the Amazons in Greek mythology.

Photos Dean Smith

12 *Voyage to Australia & the Pacific 1791-93 (Edward & Marise Duyker) Note 21 p340*

The Monument Bruny Island SE

The south east coast of Tasmania is noted for its spectacular sea stacks. The Monument is a descriptive name for a remarkable sea stack that bears a striking resemblance to an obelisk or monument. Perhaps this natural monument unwittingly commemorates the many early navigators who visited Adventure Bay for water supplies in the 18th century, including Furneaux 1773, Cook 1777 and Bligh in 1788 and 1792.

Photo courtesy of Bruny Island Charters

Cradle Mountain

Cradle Mt, Lake St Clair National Park NW

If this cradle's rockin', don't bother knockin', get the hell out of there, or down will come baby cradle and all. The mountain is Tasmania's 5th highest peak at 1545m.

The outline of the mountain with its distinctive cradle between two towering peaks is probably the most photographed peak in Australia after Uluru. The mountain was named Cradle Mountain by Joseph Fossey in 1828.

It had been previously known to Van Diemen's Land Company surveyors as Ribbed Rock, which also describes the mountain very well. The first recorded European to ascend the mountain was Henry Hellyer in 1831 although some people believe that he was preceded by Fossy when he named it.

Photo Craig Hoehne

Shipstern Bluff near Port Arthur, Tasman Peninsula SE

This award-winning photograph taken in 2006 reveals the awesome power of the huge surf with the breath-taking backdrop and rugged grandeur of the aptly-named Shipstern Bluff which towers two hundred metres above the waterline. The bluff bears a striking resemblance to the high, tapering stern of an 18th century sailing ship.

Forget Bells Beach, forget Bali, forget Surfers Paradise; 'Shippies or Fluffies' (as it is affectionately termed) generates the largest waves in the Southern Hemisphere and has become a Mecca for surfers. One surfing Magazine has voted 'Shippies' as the 'heaviest wave in the world' and is only rivalled by Tahiti and Hawaii. Only the cream of surfers operate here, as this exposed reef break is a dangerous place to surf. The hazards include rips, rocks, sharks and isolation. If you get injured here, there is no ship's surgeon on the Shipstern and there are no nearby hospitals or medical practitioners to put a dumped Humpty Dumpty together again.

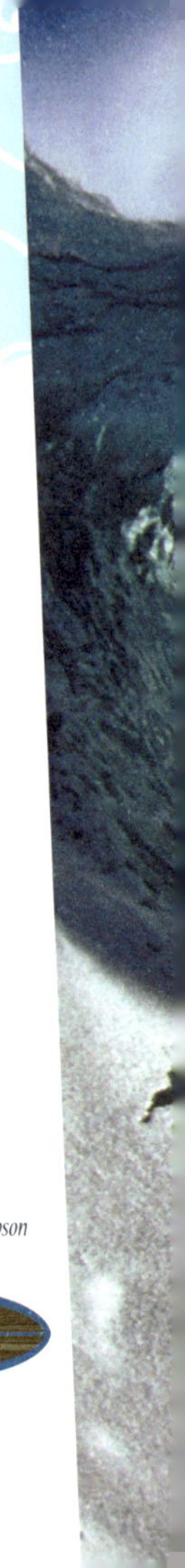

Photo Stuart Gibson

Oodles of Boos, Coos & Doos

Tongataboo District

Weetah via Deloraine NT

There is nothing taboo about using this Tongan name in Tasmania, except that we never know how to spell it correctly. The district of Tongataboo (renamed Weetah in 1912) was named after nearby Mount Tongataboo, which in turn was named after the Tongatapu Islands in the Tonga Group (possibly because of its shape).

NB. Apart from Tongataboo District, there is a Tongatabu Mountain near Geeveston and a Tongatabu Road at Mt Dromedary near Boyer. It is not known why so many Tasmanian features were named after a Tongan island unless all of them resemble the island's shape. In Polynesia, *Tonga* means 'southern' and *tapu* means 'taboo' or 'sacred'.

Cooee Burnie NW

This Burnie suburb is sited on Cooee Creek, which was named by employees of the VDL Company in the late 1820s.[13] It is not known why the creek received this evocative name as it is an Aboriginal greeting or warning. It seems probable the surveyors who named the creek either cooeed to gain somebody's attention or perhaps they heard some Aboriginals calling to each other here. Cooee is the shortest place name in Tasmania featuring two double vowels.

Photo Dean Smith

13 Henry Hellyer Map 1828

Coot Shore

north shore of Lagoon of Islands, Shannon CH

Watch out ladies, there may be a randy old coot residing hereabouts. Seriously, the name is probably a reference to coots (aquatic birds) or bandicoots. (Perhaps I should amend the first sentence to read 'bandy old coot').

Cooters Layout a clearing on

Ragged Jack, Ben Lomond (Mountain) NT

This area is named after another old coot who was a coalminer with the nickname of 'Cooter' who had a coal mine on the top of Ben Lomond. He sold coal to be used for heating to the occupiers of huts on the ski field.[14]

Kalangadoo

Lake Leake, Northern Midlands OC

This fishing village is sited at Kalangadoo Bay on Lake Leake. The bay that was created when this area was flooded contained some large gum trees and because of this, the bay was subsequently named after the town of Kalangadoo, South Australia. Kalangadoo is a SA Aboriginal word meaning 'red gums among the swamp'.[15]

14 *Tasmanian Place Names* (Taylor & Smith)

15 Source - Sign at Kalangadoo Store, Council publication *Lake Leake*

Doo Town Eaglehawk Neck, Tasman Peninsula SE

Doo Drop Inn to this interesting fishing village on Pirates Bay as it is visited regularly by tourists because of the creatively named shacks and houses here. Each one features the word '**Doo**' in its name. This theme began in 1935 when Eric Round, a Hobart architect, named his fishing shack here *Doo I*. Soon he was joined by his friend Charles Gibson who named his shack *Doo Me* and many others followed.[16] Some examples of shack names here today are shown below. There is one dissenting house in the town, daringly named *Medhurst*.

16 Nomenclature Board records, personal knowledge, http://en.wikipedia.org/wiki/Doo_Town,_Tasmania http://travelmedia.tourismtasmania.com.au/about/quirkytas/places.html

Pious Addresses

Tasmania is fortunate to have several groups of Biblical place names which create a lot of interest among visitors. Tourists often pose for photographs in front of town signs like Paradise, Promised Land or Garden of Eden which are all located in the Sheffield district. These towns were named as follows:

Paradise near Sheffield NW

This was first recorded as 'Reuben Austin's Paradise' and was eventually shortened to Paradise. This is an appropriate name as the settlement is established in gorgeous rolling countryside facing the spectacular bulk of Mt. Roland. As the area developed, other settlements were given Biblical names by Christian Brethren to maintain the theme; e.g. Beulah (Biblical name for Palestine), Garden of Eden, and Promised Land.[17]

NB. Blesseds Shore (mentioned on the back cover) is located on Lake Echo in the Central highlands. It was not a Biblical name at all, as it was named after a local shepherd by the name of 'Blessed' Bessell. There is another cluster of Biblical names in the Walls of Jerusalem National Park in the North-West. Goshen and Priory near St. Helens on the North-East Coast are also Biblical names. Neither of these clusters is covered in this book.

'Hellfire in Paradise' *Photos Craig Hoehne*

17 Nomenclature Board records

Blessington

Ben Lomond (Mountain) Nt

Photo Peter Daalder

The blessed residents of sleepy Blessington are blissfully unaware that their town was named after one of the most scandalous women in the Old Country. The town was named c1841 after Irish-born novelist the Countess of Blessington, Marguerite Gardiner (nee Power) (1789-1849). The countess was the sister of Tasmania's Surveyor General Capt. Robert Power (1794-1869) who had the final say on nomenclature up until the 1850s.[18] Countess Blessington was a woman of rare wit and beauty who was affectionately described as 'the most gorgeous Lady Blessington'. She was a generous and extravagant socialite who became a mistress to several men and was also involved in many juicy scandals including an affair with the equally salacious Lord Byron. She was also 'friends' with Dickens, Thackeray, Disraeli and many other influential men. After her aged husband's death in 1829 she stunned the British social set by running off with her much younger son-in-law Count D'Orsay who deserted his wife to live with her. He was 21 and she was 33 at the time. Despite the scandal, they remained together until she died.[19] The name Blessington derives from Blessington, County Wicklow, Ireland. This town's gaelic name was Baile Conaoine 'Townland of Favour' but became confused with Baile Conaoin 'Holy Communion' later anglicised as 'Town of blessing'.[20]

18 AOT – Pretyman Index

19 Literary Encyclopedia (http://www.litencyc.com/php/speople.php?rec=true&UID=1679)

20 *Tasmanian Place Names* (Taylor & Smith)

More Biblical Names

Prepare yourselves for a ripper yarn. Many southern midland townships were reputedly given Biblical names around c1806, by a party of convicts led by Royal Marine Private Hugh Germain who were hunting kangaroos and emus for government supplies. This story was first recorded by Quaker missionary James Backhouse in 1834 who asserted that:

'Only one of the party could read; and his only books were a *Bible* and the *Arabian Nights*… out of which books the places were successively named…'. Backhouse then specifically mentions Jericho, Bagdad and Abyssinia as examples.[21]

However, this story is questionable on several grounds, namely:

- It was based on anecdotal evidence from Germain about events that took place three decades beforehand. Even if Germain was not promoting himself (as I suspect he was), the truth is often exaggerated and embellished over a long period of time.
- Abyssinia was not named by Germain at all. Scottish emigrant John McRa named his land grant at Bothwell 'Abyssinia' in 1825, reputedly 'as it was an Aboriginal hunting ground'. Black Tier is nearby.[22] The place name 'Abyssinia' did not appear on a map until 1826.
- It would appear to be logistically impossible for Germain to cart fresh meat in 1806 from far-flung bush wilderness areas such as Jericho and Abyssinia, as there were no roads. A heavily laden bullock dray would take the best part of a week to travel to Hobart and by that time the meat would not be suitable for human consumption. In addition, the Rev. Knopwood's servants were catching kangaroos easily in and around Hobart in 1806. This begs the question why would Germain need to hunt in the Midlands?
- The jury is still out on this wonderful but implausible story.

In my opinion, Jericho and Bagdad were named in association with the nearby Jordan River (which was probably named by a government surveyor).

Since the Backhouse Journals recorded Germain's exploits, the story has been embellished considerably and any place with a Biblical name has been attributed to Germain. It has been suggested that the ubiquitous convict Jorgen Jorgenson accompanied Germain (but he was overseas at this time). Germain supposedly served in Egypt and then named places in the midlands after battle grounds in the Middle East. Although he was illiterate, Germain was reputed to have read stories from the *Bible* and the *Arabian Nights* to local shepherds. All of these stories cannot be verified.

21 *A narrative of a visit to the Australian Colonies* by James Backhouse p212/3, *Extracts from the letters of James Backhouse* Volume 1 Part 2 p36

22 *Bothwell The Gateway to the Highlands* (Gwen Webb) p26, Map of the Settled Districts of VDL 1826-28

'SORRY SIR, THE ONLY WEAPONS OF MASS DESTRUCTION WE HAVE HERE ARE RABBITS!'
BAGDAD

Ripsnorters & Porkies

Squeaking Point Port Sorell NC

A squeaking point? Pig's bottom! You must be joking! But jokes aside, there is probably a very simple explanation for this intriguing name, as the point is very sandy and the sand probably squeaks underfoot. The stories that the point was named after an escaping piglet or a Constable Squeaker cannot be verified and must be considered as amusing, whimsical and fanciful stories.[23]

Squeaky Point a patch of bush on Smilers Spring Creek near Swansea EC

Now this story you can believe. This squeaky point was named from an escaped pregnant sow.

The Grunter Mole Creek NT

Goodness gracious! A grunting hill! What next? Seriously though, the hill **does** grunt as water thundering and echoing through Grunter Swallet (a cave system beneath this hill) sounds like a giant grunting, hence the intriguing name. A 'swallet' is an abyss (cave) into which a stream disappears.[24]

Pigsties Bay Recherche Bay SE

It is getting worse! First we have two squeaking points and a grunting hill and now we have a pig sty in a bay. Pigs might fly, but how well can they swim? Well once again, the name is misleading. In 1838 Lady Jane Franklin (wife of Governor Sir John Franklin) referred to a muddy bay here in her diaries as 'Pigsty Bay' and to the mudflats as 'Pigsties'. I don't want to be a stick in the mud, but I don't think pigs could either fly or swim if their feet were stuck in the mud at the Pigsties.

Porky Loorana, W King Is. BS

Now you might have thought that I was telling you porkies with the origins of the above weird names, but the explanation of this porky is more believable. The settlement of Porky (since renamed Loorana) was so named because of the prevalence of porcupines (more properly called echidnas or spiny anteaters), which frequently provided sustenance to early sealers.

23 *A History of Port Sorell* p28
24 Nomenclature Board records (article *The Chudleigh Lakes in Winter* by S Spurling Junior)

Eggs & Bacon Bay Port Cygnet 59

Despite the often quoted local myth that Lady Franklin, the governor's wife, breakfasted here, this bay was named after a yellow and brown pea-shaped wildflower *Dillwynia glaberrima*, commonly called 'Eggs & Bacon'. The plant grows prolifically here but is rarely seen at neighbouring bays.[25]

Dillwynia glaberrima
Photo Peter Graham

'Pig Village' Kelvedon Hills near Swansea 60

This is an unofficial name that was applied to a thick patch of bush into which a big sow escaped and had a succession of litters.[26]

25 Local resident Ed Penney
26 Nomenclature Board records (information supplied by Mrs EF Cotton, Kelvedon)

BIG Ones

Victoria has the **Big Mallee Bull** at Swan Hill; Northern Territory has the **Big Boxing Croc** at Humpty Doo (what a great name); ACT has the **Big Mushroom** at Belconnen; South Australia has the **Big Galah** at Kimba; West Australia has the **Giant Ram** at Wagin; NSW has the **Big Prawn** at Ballina (but Dunedoo NSW missed a great opportunity to have the **Big Dunny**); and Queensland has the **Big Cane Toad** at Sarina but it is better known for the **Big Pineapple** at Gympie.[27]

However, Tasmania has the **Big Apple** at Spreyton, the **Big Tasmanian Devil** at Mole Creek, the **Big Penguin** at Penguin and best of all – wait for it… (see opposite page)

Humpty Doo Croc NT
Photo Dean Smith

Big Devil
Photo Craig Hoehne

The Big Penguin
Photo Wayne Smith

Big Penguin inpersonating Santa
Photo Peter Daalder

27 http://en.wikipedia.org/wiki/Australia's_Big_Things#A.C.T.

…Tassie has the

Eugenana*

(very bad pun).

Eugenana is located south west of **Spreyton** near Devonport.

*Eugenana is a Tasmanian Aboriginal word meaning 'eagle hawk'.

Size Does Matter

The longest recorded place name in Australia is the Tasmanian name shown on page 38 which comprises 37 letters. There are also five other Tasmanian Aboriginal place names that have 27, 28, 29, 29 and 30 letters respectively.[28] By comparison, (according to the *Gazetteer of Australia*) the longest official mainland Australian place name is a hill in South Australia with an unknown origin that is called 'Mamungkukumpurangkuntjunya' (26 letters).[29] Therefore, Tasmania has six unofficial place names longer than the longest official Australian name.[30] There is a hill in far north-east South Australia sited near the Birdsville Track that once had a longer name than the SA hill recorded above. Survey maps recorded the hill as Mulka Nindra Coora Coora Taraninna (30 letters), but the long name was recently abbreviated to Mulka.[31] As a matter of interest, there is also a homestead in South Australia called 'Didyabringyagrogalong'. My response to that property name would be: 'Nowaymateyou'ddrinkallmegrogandthenchuckmeout'.

The longest names introduced by Europeans in Tasmania are shared between Bullocks Hunting Ground, D'Entrecasteaux Channel and Waddamana Power Station (all with 21 letters) and the longest single-word names are Bishopsbourne (near Bracknell) and Ballamacargie (near Forth), which both have 13 letters.

NB. The above Australian names pale into insignificance when compared to the longest place name in the world which is the full name of Bangkok, Thailand and comprises 163 letters, namely: **Krungthepmahanakornamornratanakosinmahintarayutthayamahadilokphopnopparatrajathaniburiromudomrajaniwesmahasatharnamornphimarnavatarnsathitsakkattiyavisanukamprasit.** However, most Thais simply refer to it as Krung Thep, 'City of Angels.'[32]

28 *A Study of the Palawa (Tasmanian Aboriginal) Place Names* (JA Taylor) pps 11, 18, 29, 34

29 http://www.ga.gov.au/news/archive/2003/march/

30 http://kaos.environment.gov.au/search.php?query=Western+Australia&s=&theme=all&p=3437

31 *Place Names of South Australia* (GH Manning) p215

32 http://www.thailandlife.com/ericshackle/placename.html

Big Tree Walk off Arve Loop Road, Geeveston, Lower Huon SE

This is one place everyone should see. Nowhere else in the world, will you see a big tree walk. If the tree can walk, I wonder if it could run? If so, we could enter it in the tree-legged race (or maybe the pole vault).

On a serious note, this is a short walk to a viewing area of a giant 87m tall swamp gum (*Eucalyptus regnans*) known as 'The Big Tree' that had been saved from logging. It is estimated to be 350 years old.[33]

NB. There are several Big Tree Walks in Tasmania.

33 Forestry Tasmania brochure

Longest Place Name in Australia

As mentioned on page 36, the longest place name in Australia by a long, long way is the Tasmanian Aboriginal name for the Little Swanport River. The river flows into Little Swanport on the east coast and was recorded as **TEE.BE.LEB.BER.RER.MEN.NAPE.BONE.YER.MEN.NAN.YER*** (37 letters). It is believed to mean 'a moving stream, flowing into a large estuary surrounded by hills'.[34] This is a very accurate picture of the topography of Little Swanport, which is sited in a natural amphitheatre. Very few people know about this fantastic name as it is 'hidden' among old Aboriginal place names that are no longer used.

This is unfortunate as it could be a tourist bonanza.

NB. This name was recorded by the Aboriginal Conciliator George A Robinson in 1831. The last three syllables of Robinson's name have been amended from the original 'man.man.yer.' to 'men.nan.yer', which is the conventional Aboriginal pronunciation of east coast rivers. The native names for Kenmare Rvt, Douglas River, Prosser River, Clyde River and Wye River all conclude with men.nan.yer.[35]

34 *Tasmanian Place Names – the Aboriginal Connection* (JA Taylor) p111

35 The name was first recorded 11/1/1831 by George Augustus Robinson, refer page 312 *'Friendly Mission'* (NJP Plomley)

Notes on Tasmania's Long Names

D'Entrecasteaux Channel off Bruny Island
Photo Dean Smith

D'Entrecasteaux Channel

(pronounced don-tre-cass-toe) W Bruny Is. SE

This tongue-twisting name was given by its discoverer the French navigator Admiral Joseph-Antoine-Raymond Bruny d'Entrecasteaux (1737-93).

NB. D'Entrecasteaux commanded a scientific expedition to SE Tasmania in 1792/3 and accidentally discovered the Channel through a navigational error. His name means 'of the [village of] Entrecasteaux', Aix-en-Provence, France. D'Entrecasteaux's family lived in a wonderful castle in Entrecasteaux which dominates the village.[36]

Waddamana Power Station **(pronounced wodda-mar-na) Ouse River CH**

The Hydro Electric Commission (HEC) adopted the name of Professor Alexander McAulay's highland property *Waddamana*, for their Hydro village built near the Ouse River in 1916. The property name was derived from Waddamana, a Tasmanian Aboriginal word meaning 'Big River'. A second station of this name was built here in the 1940s. The station name was originally spelt *Waddamanna*. Big River was also the early name of Ouse River (pronounced 'ooze') until it was officially named Ouse in 1820. The much feared Big River tribe of Aboriginals was based here in the Ouse River area.[37]

Bishopsbourne **(pronounced bishops-burn) W Longford NT**

Bishop Nixon purchased 'Vron' Estate here c1843 and renamed it 'Bishopsbourne' after Bishopsbourne, Kent, England which means 'bishop's creek'.[38] Christs College (1846-56) was sited here after a bequest from Governor Sir John Franklin and his wife Lady Jane c1843. The college was designed to become an Oxbridge-style institution providing university education. The venture failed and by 1877 the buildings were in decay.

Ballamacargie **(pronounced balla-mc-cargy) Don-Forth NW**

This was the pet name of a farming area on Dr. Casey's 3,000 acre '*Forthside*' property used by tenant farmers from Ireland.[39] The farmers named the area after Ballymacarbery (note spelling), Waterford, in their native Ireland. The Irish origin of the name is 'Town-land of the son of Cairbre'.

36 Nomenclature Board records
37 *Lifeblood* (R Lupton) p53
38 Nomenclature Board records
39 http://www.tasmaniacentral.tas.gov.au/site/page.cfm?u=259

Bullocks Hunting Ground near Targa Nt

No bull – this is one of Tasmania's horniest dilemmas (or should that be 'thorniest?'). It is arguably our most disputed place name origin. There are many explanations given, but the most logical ones are:

- Surveyor James Calder mentioned a search for surveyors' bullocks near here. Surveyors usually used bullock drays to carry their food, supplies and equipment during long surveys. I favour this explanation as being the most likely. [40]
- Nomenclature expert John Taylor claims that a man named Bullock was employed by the Launceston Military Garrison to snare wild animals for meat. This may have been George Bullock, a convict who eventully settled in the St Marys/Fingal area.
- Leslie Norman believed that the area was named after an early grantee named Bullock (which may be the same man as above).
- The name could refer to escaped cattle from the Government Stockyards located near Corra Linn, North Esk. The journal of NSW Governor Lachlan Macquarie 8/12/1811 mentioned that 613 head of horned cattle were kept at the stockyards. Escapees may have been 'hunted' down at Bullocks Hunting Ground which is not too far distant.

NB. The area became famous when the notorious bushranger Matthew Brady was captured here in 1826 by John Batman who later founded Melbourne in 1835. A few months after capture, Brady was executed by hanging. The correct spelling of his name was Bready.

40 AOT MS88 p67, P50

Shortest Tasmanian Place Names

Don River Devonport NW

This was originally called The Third Western River by Flinders in 1798. It was probably renamed by Edward Curr of the Van Diemen's Land Company after its namesake which flows through Curr's birthplace in Sheffield UK.[41] Don is derived from Br *dana* = 'water'. The river was also shown on an early map as *Musk Tree River*.

Dee River South of Lake Echo CH

Formerly Dick Browns River. It was renamed by Governor Sorell c1820 after the Scottish River Dee,[42] the name of which means 'holy river' (derived from *Deva* Gaelic for 'the goddess').

Ayr now Trenah, South of Ringarooma NE

The town was almost certainly named after 'Toon o' Ayr' by Scottish bard Robbie Burns. Ayr in Scotland literally means 'the mouth/place of the river'. The town was renamed Trenah in 1922. There is also Lake Ayr near Lake Rowallan CH.

The shortest Australian place name is a hill called Me (NSW).[43] Mainland Australia has many three-letter town names including Ayr (Qld), Cue (WA), Dum (isn't this a great name?) (NSW), Emu (Vic), Esk (Qld), Hay (NSW & Wa), Kew (NSW & Vic), Lue (don't spell it loo) (NSW), Mil (NSW), Moe (Vic), Rye (NSW), Uki (WA), Yea (NSW) and Yeo (NSW).

In a similar vein Tasmania has a plethora of rivers which also may qualify among Australia's shortest river names. These include Arm, Cam, Dee, Dip, Don, Emu, New, Old, Pet, Que, Ray, Wey and Wye.

41 Nomenclature Board records

42 Nomenclature Board records

43 http://www.ga.gov.au/news/archive/2003/march/

Den Cave Mole Creek Nt

According to Webster, a den is defined as: '*a private sanctum, a cave, a hollow, a disreputable haunt*.' The Den, Lake River and Den Hill near Bothwell were probably similarly named. Several bushrangers frequented these areas in the early 1800s and the names may have been coined to denote them as a bushranger's 'disreputable' haunt. The term is derived from OE *denn* = 'a cave'.

Limestone formations, King Solomons Cave, Mole Creek.
Photo Peter Daalder

Shortest Australian Street Name

My Street

East Launceston Nt

Imagine the reaction on the other end of the telephone line when someone asks 'Where do you live?' Answer – 'My Street.' 'Yes, but where do you live?' It would take a while to sort it all out. This is almost certainly Australia's shortest street name. It was named in 1857 by prolific subdivider James Bennell (1809-1876). According to his grand daughter Helen Davies,[44] Bennell was showing a friend over his subdivision when he was asked who built the street. Bennell replied, 'This is my street'. From this chance remark, Bennell was motivated to name the narrow laneway 'My Street'.

It is widely believed that this street was named after an altercation between Bennell and the Launceston Council.[45] However, there is no correspondence to support this view.

44 *For the Record* (Helen Davies) p172-3
45 *History in our Streets* (J & D Morris)

Rude Aboriginal Place Names

The downside of Europeans applying Aboriginal words as place names, is that the original meaning was often misunderstood or misinterpreted. Sometimes the same word meant several different things in other dialects. For example: Legana was chosen as a name c1910 because it meant 'freshwater' in one Aboriginal dialect. This was appropriate as the town is located near Fresh Water Point on the Tamar River. However, in another dialect, Legana means 'to excrete', so Leganians should feel lucky that their town name was based on 'fresh water' instead of 'fresh poo'. This example reveals that towns were given Aboriginal words as place names in the early 1900s, with only scant information about the various Aboriginal languages, and this has led to some less than desirable name applications (see list below).[46]

Mattah	= testicles' or 'death'	**Tanina**	= 'to break wind'
Tunnack	= 'frigid'	**Cokura**	= 'jerk'
Loyatea	= 'sexual intercourse?'[47]	**Tyenna**	= 'dung/excrement' or 'bandicoot'
Kamona	= 'venom'[48]	**Kamena**	= 'chin' (**But** Kamena Mina = 'to spit')
Parrawe	= 'to cease/abstain'	**Pallawah**	= 'black man' (**But** Pallawah proina = 'fat man')
(**But** Parrawe Tiakrangana	= 'to disembowel')		

46 The place name origins (above) were sourced from Ling Roth's acclaimed work '*Aborigines of Tasmania*' page lii Appendix F Tasmanian-English Vocabulary.

47 Ling Roth has no word for Loyatea and (according to John Taylor) it appears to be a contraction of Loanga metia = sexual intercourse.

48 Sometimes for simplicity, only one of two descriptive words was used as a place name e.g. Kamona is a contraction of Kamona Moina.

LOYATEA
COKURA
MATTAH CREEK
PARRAWE
TANINA

Rude Introduced Names

Buggery Beggary Bumps, Arthur Ranges SW

Do not be surprised if you ask for directions to the Arthur Ranges as you may be told to 'go to Buggery'. This naughty name was adopted for one of the Beggary Bumps (a series of ragged peaks which are renowned for being difficult to traverse).

Boy in the Boat Rock Maria Is. EC

My father once told me about 'the birds and the bees' and the 'boy-in-the-boat' and I didn't know what he meant. I have just discovered that the latter was probably this small rock island which resembles a boat with a small figure in it.

Bloomer Creek Cranbrook EC

Did someone lose their bloomers here? I am more inclined to think that the creek flowed through an area of wildflowers.

Cox Bight South West Cape SC

This painful-sounding name commemorates the visit of English Captain John Henry Cox who anchored here in 1789.[49] John Henry Cox is a fine upstanding name, don't you think? His ship was appropriately named the *Mercury*.

Clyte Bight Bathurst Channel, Port Davey SW

Now called 'Clytie Cove'. This evocative rhyming name conjures up all sorts of titillating images. However, it is a nautical name to commemorate a yacht of this name belonging to William J. Lindsay. The yacht is named after Clytie a water Nyph in Greek mythology.

Mole Creek Caves Mole Creek via Deloraine Nt

Holy Moley! Where did this name come from? There ain't no moles here, but there's holes all over the place (Mole Creek is renowned for its spectacular cave systems). The creek was originally named Mole River, after its Surrey, England namesake, which continually disappears underground (like the European burrowing mole) and reappears further downstream. The Tasmanian stream received its strange name for the same reason.

Belchers Lookout National Park St

Burping is prohibited here. The lookout was named for William ('Will' or 'Bill') Belcher, first Park Ranger at National Park 1917-34.[50] The name Belcher is derived from several sources: e.g. 'one who belches', 'good sire' or 'grandfather', 'dweller at the ridge' or 'nice face'.

49 Nomenclature Board records

50 Nomenclature Board records

BURP
BELCHERS LOOKOUT

Politically Incorrect Names

Abo Creek Strahan WC

This creek was named to commemorate the fact that the area was the domain of Aboriginals until the 1830s. The word Abo is now regarded as an offensive term. You wouldn't get away with applying this name in today's politically correct climate.

Dago Plains SE of Sandy Cape NW and Dago Point Interlaken CH

This derogative name records that an Italian had a camp in both of these places.[51] The offensive term **dago** refers to Italians and Italian-Australians, especially recent immigrants and others not fully assimilated into Australian culture.

NB. The term 'dago' is a bastardisation of the name "Diego". It was formerly used to describe Spanish sailors by British naval personnel during the 18th and 19th centuries.

Chink Point Bathurst Harbour, Port Davey SW

At first glance it may be construed that this point denigrates Asians. However, it is a descriptive name indicating that the point has a chink or indentation in it.

Blackboys Opening (Pass) North of Ouse, Lyell Highway CH

This name may sound rather rude but it is just the common name for the once prolific Grass Tree plant (*Xanthorrhoea australis*) which is also popularly known in Tasmania as 'Kangaroo Tail'.

Black Gins Point Great Lake CH

Named after an unidentified female Aborigine.

Tittie Gee Creek Emu River, S Burnie NW

Gee, this mammorial memorial probably commemorates a Chinese prospector who worked this creek.

Tonguers Point Port Davey SW

With tongue in cheek I will admit that this seemingly rude name commemorates the men known as 'Tonguers' who tried-out (boiled down) whale blubber on behalf of whaling crews during the 1840s and 50s. Tonguers get this strange name because they cut out the tongues of the whales.[52]

51 Nomenclature Board records

52 Tasmanian Tramp No.14 p47

Pisspot Creek Little Swanport EC

It is uncertain whether this pissy creek was named after a chamber pot found here or a habitual drunkard living nearby. Although it is purely speculation, I suspect that the creek may have been named after Paddy de Courcey Harte, who was a well-credentialled but dissolute identity of Little Swanport in the 1830s.

Bungs Point

Cape Barren Island, Furneaux Group BS

This point is not bung at all, it is named after 'Bung', a seemingly insulting English nickname given to an Aboriginal woman called Wore.ter.moe.te.yen.ner (aka Ware.moe.deen.ner), who was abducted by sealer George Briggs (with whom she had three children).

NB. Briggs sold her to sealer 'Long Tom' Thomas for a guinea. She later lived with sealers Brown and Munro. At one stage, she also had an Aboriginal 'husband'. To have kept this many husbands happy, 'Bung' could hardly have been bung.[53]

Kafoozalum Creek

East of Tooms Lake EC

The late Phil Bowling of the former Forestry Commission suggested this cheeky name from a bawdy campfire song *'Kafoozalum, the Harlot of Jerusalem'*, because of 'difficulties' Forestry workers experienced in the area. One would assume that foresters found that 'life's a bitch' here.[54]

Niggers Head Strahan WC

The politically incorrect name of this headland probably commemorates local Aboriginals who lived in this area prior to the 1830s. However, not all 'nigger' names are demeaning. There is a Niggers Lane in Kingston that commemorates the nickname of a local white resident.

Upper Plenty

Feilton near New Norfolk SE

Despite the raunchy name, this was a town sited on the upper reaches of the Plenty River in the 'Valley of Love' (Derwent Valley). Political correctness saw this titillating name changed to the blander Feilton. Why do do-gooders always have to take the fun out of everything?

Punk Plain NE of Guildford Junction WC

I don't find punk 'plain' at all, as punk music is deafening with raucous and obnoxious lyrics. However, we are talking about a different kind of punk here. Punk is the common name for a fungus that was used as tinder. It was also reputedly eaten by Aboriginals.[55]

Red Rag Scarp

Wellington Range, South of Hobart SE

Rag is a geological term for pebbles or small rocks. Obviously these rocks were reddish in colour. A scarp (escarpment) is a line of cliffs usually caused by faulting. In other words, a rock fracture usually caused by earth movements.[56]

---

53 *Friendly Mission* (NJP Plomley)
54 Mr Chipman advised the Nomenclature Board 21/11/1979
55 *Tasmanian Place Names* (Taylor & Smith)
56 *Glossary of Geology* (American Geological Inst.)

Black Charlies Opening Runnymede, Tasman Highway

The original name for this opening was Parsons Pass as it was sited near 1560 acres of land known as *Brush Farm* granted in 1831 to local farmer Charles Octavius Parsons JP (1799-1863), who had a magnificent jet-black beard. It seems fairly obvious that Parsons was nicknamed 'Black Charlie'.[57]

Because of the evocative name, there have been some very inventive and fanciful stories as to how the pass acquired its appellation. According to whatever myth you chose to believe, Black Charlie was (in order of popularity):

- An Aboriginal (or a swarthy white guide, probably Charlie Parsons?) who assisted road surveyors to find a way through the mountains to the East Coast in the 1820s.
- Charles 'Black Charlie' O'Connor a Sorell farmer of Indian extraction. He would have only been 12 years old when the name 'Black Charlie' was first recorded (1831).
- A black servant of Sorell magistrate James Gordon who was reputedly murdered here whilst on the way to deliver a message. His boots were allegedly hung up on a tree by the road and remained there for many years. Why an Aboriginal would require boots has never been explained.
- A black-tracker chasing bushrangers.
- Any number of other local Charlies (who were mostly born after the name became popular).

'Black Charlies Farm' was first mentioned in the journals of Aboriginal conciliator GA Robinson in 1831.[58] As Aboriginals did not 'farm' (in the European sense of the word), this eliminates all claims that the pass was named after an indigenous person. Once again, the myths are much more 'interesting' than the real origin.

57 Parsons came from Monmouthshire in Wales per *Princess Charlotte* arriving in January 1823 and became a storekeeper at Hells Gates, Macquarie Harbour and later he served as a Commissary Clerk on Maria Island penal settlement c1827. This was probably the time he began farming at Runnymede.

58 Sources: Richmond historian Alex Green; *Friendly Mission* (Plomley) 16/1/1831; page 316, AOT County Charts - Pembroke 2 and 2A; ADB.

Forcett via Sorell SO

This name calls for another shocking pun and I am about to 'force it' onto you.

There was a fat lady of Forcett,

who couldn't fit into her corset.

She said to her maid 'It's too small I'm afraid.'

'Twill fit,' said the maid, 'coz I'll force it'.

This small township derived its unusual name from Magistrate James Gordon's 243ha land grant of 'Forcett' nearby. Gordon was born at Forcett, Yorkshire, England (which probably means 'fold by a ford'). Gordon became the founder of the township of Sorell in the early 1820s.[59]

59 Nomenclature Board records

Colourful Characters

Swearing Bobs Plains Waterloo, lower Huon SE

This @#*›@# name refers to an intriguing character who habitually used colourful language. Apparently he had to use at least one swear word in every sentence. 'Swearing Bob' was none other than Robert 'Bob' Ratcliffe, a bushman who worked for Geeveston saw-miller Osborne Geeves in the 1880s. Apart from his swearing, Bob was renowned for his skills as the maker and sharpener of timber-splitting knives. Unfortunately, timber splitting is now a forgotten craft.[60]

Jack the Liars Creek Fingal EC

I would be lying if I pretended that I didn't like this wonderful and evocative name as it is another example of our colourful past. As this creek is located in an old mining area, we can be relatively sure that this place name relates to someone who lied about a mining strike.

Laughing Jack Lagoon Derwent Bridge CH

Getting to the bottom of this place name is no laughing matter as the origin is not as simple as it sounds. The intriguing name seems to commemorate the nickname of a jolly pioneer (a shepherd perhaps?) who was habitually laughing. It may also commemorate a settler's mule or jackass which may have been located in this vicinity. The name was in use before the introduction to Tasmania of laughing jackasses (kookaburras) in 1877.

Molly Yorks Nightcap Western Tiers near Lake Sorell CH

The story behind this place name is so cute and believable that it is almost certainly a fabrication. Molly York was reputed to be an unruly assigned servant (convict) who missed serving breakfast one morning. Her excuse for her absence without leave was that she had been climbing the mountain peak that is now named after her. She was not believed until her nightcap was found beneath a rock on the outcrop.[61] As a matter of interest, there were no female convicts transported with the surname of York.[62]

60 *Heaven is our Home* (Lance Geeves)
61 Nomenclature Board records
62 AOT convict index

Flash Charlies Marsh

Great Western Tiers CH

This wonderful place name commemorates a man who wore gold sovereigns for buttons on his waistcoat.[63]

63 Nomenclature Board records

Crackpot Caves

Mole Creek Area via Deloraine Nt

Tasmania is fortunate to have many cave systems and each cave is usually given a clever, fascinating or fanciful name. Although cave names have to be registered they have been largely free of the restrictions applied by naming authorities that tend to frown on 'frivolous' names. In other words, cave names have not been 'dumbed-down' to conform to conservative naming policies.

Definition: Pot: … a general term for any hole, pit or depression produced naturally in the ground and often containing water that suggest the shape of a large (cooking) pot or boiler.[64]

Crack Pot
Long Shot Pot
Rabbit Trap
Swiss Cheese Cave
Toboggan Cave
Devils Earhole
Spider-Pyramid-Cow System
Kubla Khan & Ghenghis Khan [66]
Flowers Pot [68]
Execution Pot
Aladdins Cave
Shish Kebab
T Bone Cave
Lillians Rift
Sharlands Organ
Croesus [65]
King Solomons Cave [67]

64 Glossary of Geology – American Geological Institute

65 The term as 'rich as Croesus' (pronounced Cree-sus) comes from the legendary Croesus, King of Lydia, Turkey c560BC whose name became synonymous with wealth. The name was used to emphasise the rich cave formations found here.

66 The name derives from the fabulous diamond of that name as this is a fabulous cave with 'rich formations'. The diamond was named after Kubla Khan (1215-94), a grandson of Ghenghis Khan, the first Mongol emperor of China and the acknowledged head of the Mongolian hegemony which stretched to and over Persia and Russia.

67 The allusion is to the fabulously wealthy Biblical king of the Jews.

68 A clever name that commemorates the Flowers family who were pioneers at Mole Creek.

Iron Pot (lighthouse), Derwent Mouth

The Iron Pot Photo Dean Smith

This potty name was applied to a small rocky island on which it is believed an iron whaler's pot was left in the very early days of settlement. The island was referred to by this name by adventurer and whaler Captain James Kelly on 30/1/1816 on his return from circumnavigating Van Diemen's Land. This was well before the lighthouse was built c1832.

Whalers frequently established whaling stations on rocky shores where whales could be dragged up onto a convenient rock ledge to be cut up and boiled down in cast-iron whaling pots. There have been unverified stories that a whaler's iron pot was located here with a fire in it to serve as a beacon to warn approaching ships of impending danger.[69] The light has been renamed Derwent Light but everyone ignores this boring name and they affectionately retain the more evocative 'Iron Pot'. There seems little substance in the popular theory that the island was named from 'pot-holes' in the rock surfaces.

69 Nomenclature Board records

Fruity and Nutty

Pearshape District King Island BS

What's gone pear-shaped here? Well that's easy to answer. The nearby Pearshape Lagoon is in fact pear-shaped and the district was named after it.

Crabtree Huon SE

A crab tree? You've got to be kidding. I've seen trees with saplings and suckers but never one with little nippers. We all know that money doesn't grow on trees and I suspect that crabs don't grow on trees either. Well, I'm only pulling your leg, because crabs really **do** grow on trees. A French Crab is an old variety of apple that is usually only suitable for cooking. The Crabtree district was established around Crabtree Farm owned by Charles and Mimosa Parsons c1860. The property reputedly gained its name from an old crab apple tree which was growing there when Parsons purchased the land. *Crabtree House* on the old farm is now used for colonial accommodation.

The Nut Stanley NW

Residents of Stanley should be called 'Nutters' as they live in the shadow of The Nut, an imposing rocky eminence. Just imagine getting a letter addressed to you as:

Mr. W. Smith, The Nut, Nut Reserve, STANLEY, TAS.

How could anyone take you seriously after that? This then begs the question as to which nut named The Nut? Whoever it was must have had rocks in his head. However, he wasn't as nutty as Matthew Flinders who was a proper 'fruitcake' as he called the rock Circular Head in 1798 because he thought it was: 'a cliffy, round lump, in form much resembling a Christmas cake'. Since when did a nut resemble a Christmas pudding? Many people suggest that The Nut is shaped like a walnut but I don't see the resemblance. The spectacular Nut is Tasmania's equivalent of Uluru (Ayers Rock) as it can be seen for kilometres in any direction. The Nut is the remnant of a giant volcanic plug formed during lava flows.

Photo Peter Daalder

Tassie Double-Bunger Names

Unfortunately, unlike mainland Australia, Tasmania is not blessed with many double-barrelled names like Wagga Wagga, Gin Gin, and Curl Curl etc. However, we do have a few, and these are listed below for your amusement with one ring-in to make up the numbers.

Bung Bung Creek

tributary of Little Pine River, west of Great Lake CH

Bung means 'dead' in NSW and Vic Aboriginal languages (i.e. creeks that usually flow only in wet seasons and become dry creeks in summer). In most instances, doubling an Aboriginal word means 'more than one' or 'many', so 'dead' becomes 'even deader' (although I don't know what could be deader than dead). Paradoxically, in this case, it may mean the reverse (i.e. 'big water' or 'many waterholes').[70]

Dubbil Barril Railway Station

on the ABT line, Strahan WC

This fascinating (unverified) name is almost certainly named after the nearby King River which is divided here by a long-wooded island, thus forming two streams. The station may also earn its name from the more distant Double Barrel Creek although this appears to be a mere coincidence.

Dubbil Barril Railway Station
Photo Dean Smith

70 *Aboriginal Words of Australia* (AW Reed) p74, 77; *Aboriginal Place Names* (AW Reed) p22

Mount Barra Barra (450m), Upper Isis River NM

This mountain is located on 'Barra Barra' Estate, established by John York of Auburn. Barra is a Queensland Aboriginal word for 'grubs'. By doubling the word it means 'many grubs'. Perhaps many of the early residents were 'grubs'.[71]

Mount Yopyop (629m), Olga River WC

This mountain was named from the Russian cursing expression reputedly uttered by Lithuanian adventurer Olegas Truchanas (who wasn't a Russian and who obviously wasn't rushin') when battling to the pinnacle through particularly heavy growth. Unfortunately, this interesting name has been 'sanitised' as the peak has been renamed Olegas Bluff as a tribute to Truchanas. However, Truchanas has also been honoured in other Tasmanian features: e.g. Truchanas Huon Pine Reserve – so the original name could have prevailed.[72]

Yo - Yo Track Huon-Port Davey Track SW

This track gains its amusing name as it goes up and down (steep hills) like a yo-yo.

71 *'The Children of Erin' A History of Campbell Town* (G. Duncombe) p6; *Aboriginal Place Names* (AW Reed) p11
72 Nomenclature Board records

Cuckoo Names

Cuckoo SW Scottsdale NE

This cuckoo name was named after Cuckoo Valley in which the township is sited. The valley was named after the cuckoo bird. There are four species of cuckoos in Tasmania. That adds up to *'Cuckoo, Cuckoo, Cuckoo, Cuckoo'*.

Growling Swallet Florentine Valley SW

What on earth is a 'growling swallet'? A swallet is 'an abyss into which a stream disappears'. This particular cave is one of Australia's deepest caves and gains its strange name because the river which tumbles into it makes a deep, rumbling noise.

Look-In Lookout Arve Road, near Geeveston SE

Next time I am in Geeveston I am going to 'look out' for this lookout so that I can 'look in'. I 'looked up' 'lookout' (in my *Funk and Wagnells*) and I was advised to 'look down' at the scenery. Now I am confused. I now don't know whether to look up, down, in or out.

NB. According to Forestry Tasmania literature, you can look 'into' the forest from this elevated platform. Look out for it.

Cramps Great Lake CH

The naming of this highland fishing village has nothing to do with a medical condition. It was named after William Thomas Cramp and his family on the recommendation of the Fisheries Board in 1954. Cramp founded the Cramp Hatchery for brown trout at nearby Interlaken c1910.[73]

Flintstone Arthurs Lake CH

This collection of holiday huts was so named in 1999 as dolerite and other flinty stone abounds here.[74] The settlement is predominantly populated by anglers and fishermen and it was named as a joke (with a clever play on words) after the TV cartoon character 'Fred Flintstone'.

NB. Some shack names here reflect characters, places and sayings from the *Flintstones* TV series such as: *Rubbles Retreat, Bedrock Cottage* and *Yabba Dabba Doo.* Many of the shacks have comical names reflecting the rocky topography like, *Rock Bottom, The Rockpile, The Freestones* and other holiday themes including *Booze & Snooze, Hilton on Arthurs (Lake), Driftys, Dog Haven, Just a Shack* etc. Streets here are named after types of stone: e.g. Flintstone, Chert, Slate, Quartzite etc.

Humbug Point St Helens NE

Bah humbug! This point was not named by the prince of humbuggers (Charles Dickens's *Scrooge* from *'A Christmas Carol'*). It was reputedly named by government surveyor and farmer JH Wedge because of many sheep and lambs being lost here owing to the depredations of Tasmanian tigers.[75]

73 *Tasmanian Place Names* (Taylor & Smith)

74 Nomenclature Board records

75 Nomenclature Board records, Royal Society of Tasmania, Bulletin No 4, 1973

Hobler's Bridge Launceston NT

The amusingly named bridge commemorates George Hobler (1801-81) who arrived in VDL in 1826 and was granted land near the bridge.[76]

76 *History in our Streets* (Morris) p64

Hill-arious Hills

Going Hill Port Davey SW

Despite its name, this hill is going nowhere. It was named after Lt. Philip Going R.N. 2nd in command of HMS *Havannnah* which charted this area c1850.[77]

Hill Of Blazes River Ouse CH

Who the blazes thought of this fiery name? In reality, the name has nothing to do with forestry blazes, it was simply covered with trees, some of which contained blazes [marks] cut by Aborigines.[78]

Bread 'n Butter Hill Elliot NW

This sounds like a good place for a fast food outlet. The hill reputedly gained its name when carters found it necessary to rest their horses and bullocks after climbing the hill and ate their lunch there whilst waiting. I have great difficulty swallowing this (forgive the awful pun). I am more inclined to think that the farm (on which this hill is sited) was successful, thus providing food for the plate.[79]

Bulgobac Hill Bulgobac District WC

This reads like a warning sign on a farm gate 'BULL – GO BACK'. However, the hill was named after Bulgoback (note spelling), near Balook, Wellington Shire, Gippsland, Victoria possibly because of its resemblance to a mountain there.

NB. The unusual name probably means 'district behind Bulga' as Tarra-Bulga National Park is sited nearby. The name appears to be derived from *Bulga* = NSW Aboriginal for 'hill' + English *back* meaning 'behind/at the rear of'.[80]

Bendover Hill Portland NE

Origin unknown, but if you are wise, you do not bend over here as you may be butted by stud rams or billygoats lurking nearby. However, that is not the origin of this strange name, as the hill probably 'bends over' near the top. It is possible (but unlikely) that the name may be an intimation of the arduous and back bending occupation of mining as the hill is located in an old tin mining area.

Dreadnought Hill S of Renison Bell WC

This fearless hill was named after the Dreadnought Mine which gained its name from a famous British battleship completed in 1906 that set the style for armour-plated battleships for the next 40 years.

NB. The driving-force behind the development of the *Dreadnought* was Admiral John Fisher, the First Sea Lord. The ship was built at Portsmouth dockyard between October 1905 and December 1906. It was the most heavily-armed ship in history. She had 10 long-range 12-inch guns (305 mm), 24 three-inch guns (76mm) and five torpedo tubes below water. In the waterline section of her hull, the *Dreadnought* was armoured by plates 28cm thick. Dreadnought means 'fear nothing'.[81]

77 Personal research
78 *Tasmanian Place Names* (Taylor & Smith)
79 *Tasmanian Place Names* (Taylor & Smith)

80 Information gained from a website on Bulgoback, Victoria (since removed).
81 http://www.spartacus.schoolnet.co.uk/FWWdreadnought.htm

Batman Lookout Ben Lomond Nt

There is no sign of super-heroes 'Batman' and his friend 'Robin' here. The lookout is named after the celebrated John Batman a local pioneer farmer who later co-founded Melbourne in 1835. He achieved fame by capturing the notorious bushranger Matthew Brady (see Bullocks Hunting Ground page 36 and 40). Batman's real name was Bateman.

Phantom Peak Arthur River WC

They seek him here, they seek him there – but you won't find the super comic-book hero 'The Phantom' here (nor will you find Batman & Robin). The origin of this name is unclear but it is likely that the peak seems like 'an apparition' as it appears, disappears and then reappears with the vagaries of mountain mists (or from behind foreground hills, if the observer is moving).

Constitution Hill Midland Highway SM

The popular myth is that this steep hill gained its name as it would test one's constitution. However, this theory is fanciful guesswork, as the hill was named after the famous Constitution Hill in London which is just a gentle rise on the route taken by most royal processions from Buckingham Palace.[82]

Rattler Hill Weldborough NE

This name once had me rattled, as hills don't usually rattle (even if you shake them). Furthermore there are no rattlers (rattle snakes) in Tasmania. But there is an obvious answer, as the hill was named after Rattler Creek (which presumably 'rattles along at great speed' after it rains).

Thieving Hill Hamilton St

What on earth did the hill pinch? The mind boggles. The allegedly light-fingered hill is located in an area renowned for stock thieving and therefore stock grazing on the hill were probably subjected to duffing. This seems to be the logical answer for the suggestive name.

Ding Dong Hill Upper South Esk NE

Some ding-dong named Ding Dong Hill after the 'Ding Dong' Tin Mine in this area (but how can you call a hill 'Ding Dong' if it ain't got a bell to ding or dong?). There were many Bell Hills and Bell mines in Tasmania named after the enterprising mining prospector George Renison Bell (1840-1915). This appears to be a humorous alternative to naming yet another mine (and hill) after him.

The Battery Hill Beaconsfield, Tamar Valley Nt

This battery never went flat (until they turned the power off). It was the site of a gold ore-crushing battery which operated as part of the separation process.[83]

Spring Hill Midland Highway SM

You can climb Spring Hill in Spring (and Summer, Autumn or Winter) but it is steep so you will need quite a spring in your step. However, the name refers to a freshwater spring on the northern side of the hill which was tapped to provide a water trough for horses during the old stage coaching days.[84] They would certainly need a drink after reaching this challenging summit.

82 *Tasmanian Place Names* (Taylor & Smith)

83 *Tasmanian Place Names* (Taylor & Smith)

84 Personal knowledge

Bust Me Gall Hill

East Runnymede SE

You need a lot of gall to climb this steep hill and even then your gall may be busted. The hill was reputedly named by bullockies as bullock wagons had to toil up this steep hill with great difficulty. Frequently, passengers had to get off and walk up the hill to assist the bullocks.[85]

85 Nomenclature Board records

Break Me Neck Hill East Runnymede SE

Reputedly named by bullockies after bullock drays went over the top of Bust Me Gall Hill and carefully negotiated the equally steep descent of this adjoining hill. No doubt some wagons bolted or went out of control to cause damage and chaos.[86]

86 Nomenclature Board records

Fred & Barney

Frederick Henry Bay Storm Bay SE

Fred Flintstone from the TV series *Flintstones* would say 'Yabbadabbadoo' if he realised the comedy of errors surrounding this place name. The naming and siting of this bay was a monumental drafting error made by Arrowsmith of London in their 1832 Map of Van Diemen's Land. The cartographers believed that the Bay north of the River Derwent entrance was the site of Tasman's anchorage in 1642. However, Tasman had anchored near the present North Bay at the northern extremity of Tasman Peninsula, naming an adjacent bay (the present Blackman Bay near Dunalley) Frederick Henricx Bay (note spelling). This name honoured the Stadtholder (de facto hereditary head of state) of the Dutch Republic from 1625-47.[87] Flinders in 1798 erroneously located the bay near Tasman's anchorage calling it Frederik Hendrik's Bay (note spelling). Arrowsmith's Frederick Henry Bay was originally named Baie Du Nord by French Admiral Bruny d'Entrecasteaux in 1793 and this was anglicised to North Bay by Flinders in 1798. Ironically, this sequence of errors was compounded when the North Bay of Flinders was transferred to Tasman's anchorage. Despite all of these changes, the spelling and location of Tasman's Frederick Henricx Bay is still incorrect.

Cape Frederick Hendrick Tasman Peninsula SE

Yabbadabbadoo, number two! The drafting errors outlined above have been further compounded by the incorrect naming, spelling and siting of this cape. Cape Frederick Henry (note spelling) at Bruny Island was originally named by Capt Tobias Furneaux in 1773 who believed Adventure Bay to be the site of Tasman's Frederick Hendricx Bay (note spelling). The error was partially rectified in 1952 to commemorate the coronation of Queen Elizabeth 11 by renaming it Cape Queen Elizabeth. The cape was then resited to Tasman's anchorage on the northern extremity of Tasman Peninsula and renamed Cape Frederick Hendrick. However, this is still in the wrong vicinity (see Frederick Henry Bay above). In any event, Tasman did not name any headland after the Stadtholder of the Netherlands. Are you confused? I certainly am.

87 *Discovery of Tasmania 1642*

Barn Bluff Cradle Mountain National Park NW

Nicknamed 'Barny' by bushwalkers, this massive pillar of stone would not be an ideal site for a barn as it would be too draughty to hold barn dances here in winter and this would certainly generate a few barneys. The barn-like bluff is sited on the roof of Tasmania and is the state's third highest mountain at 1559m and it can be seen for miles in any direction. It was probably first sighted and named by Van Diemen's Land Company surveyor Fossey in 1828.[88]

Photo Craig Hoehne

88 *Explorers of Western Tasmania* (Binks) p78

Brummy Roads

Brummy Road

Manuka Road, Kettering SE

This brummy road was not a road at all (that is why it was brummy). It was a property of that name which was named after Brummy Road, Birmingham, England. 'Brummy Road' farm was established here c1857 by pioneer brick-maker Alfred Hughes who was born at Brummytown (yes, you'd better believe it), Birmingham, England.[89] Let's hope that he didn't use brummy bricks to make brummy houses here.

NB. Brummytown is the popular nickname for Birmingham. Brummie (sometimes Brummy) is a colloquial term for the inhabitants, accent and dialect of Birmingham, England, as well as being a general adjective used to denote a connection with the city, locally called Brum. The terms are all derived from Brummagem or Bromwichham, historical variants or alternatives to 'Birmingham'.[90]

Yellow Brick Road Brighton SE

This road is neither yellow, nor brick, but it **is** a road. The name was derived from the book *'The Wizard of Oz'*.[91]

NB. Prior to the application of this name, the first names suggested to the Council by the developer were Sesame Street (from the famous children's TV show) and Abbey Road (the Beatles' address) but both were refused. This exposes some of the absurd shenanigans that councils and naming authorities are subjected to from developers proposing frivolous or inappropriate names.

Smith & Others Road

Sassafras NW

This amusingly-named road leads to properties once owned by William Smith and his sons William Junior and Arthur. The 'others' residing on the road were the Lamprills and 'other' families unknown. The Lamprill family claim that they were living on this road before the Smiths arrived.[92] The Smiths were all bushmen who cut timber here in the early 1900s. Even today, this road is a forestry road.

Crooked Billet Drive Brighton SE

This crook name came from the name of a handsome old licensed public house known as *'The Crooked Billet'* owned by elderly pioneer Francis Cox in the 1830s. The house was obviously named after one of the many British pubs of this name (or less likely, from the *'Battle of Crooked Billet'* in Pennsylvania in the American Revolutionary War in 1778).[93]

NB. A billet is the place to which a person, generally a soldier, is assigned to sleep. Soldiers are generally billeted in barracks when on duty, although in some armies soldiers with families are permitted to maintain a home off-post. In this case the billet must have been a dilapidated or crooked building. A billet can also be a piece of firewood or metal.

89 *Days Gone By In The Channel (Vol 2)* (Madge Lowe) p45
90 http://en.wikipedia.org/wiki/Brummie
91 Nomenclature Board and Brighton Council records
92 Information by a listener to ABC Radio 'Nomenclature Talk-back' program
93 *Tasmanian Place Names* (Taylor & Smith)

Snotty Road near Scotch Town, South of Stanley NW

This is snot a nice place in winter and your nose will run faster than an out of control bus if you linger here. If you do, make sure that you bring a large handkerchief with you (or even a towel).
The road gets its name from a steep section on Connell Hill which is cold and slippery in winter. This area is sometimes referred to by locals as 'Snotty Nose' or just 'Snotty'.

All Time Favourite Street Names

Rue De Remarques

Lower Ballawinne Rd, Lindisfarne, Hobart SE

This private lane was given this clever double-meaning name in 1996 by local resident Warren Boyles after he heard Ronnie Barker (of the British TV comedy series *The Two Ronnies*) wisecrack that he had just returned from a holiday in Paris at 69 Rue de Remarques.

NB. As a consequence, Boyles placed a sign (spelt incorrectly as 'Rue de Remaques') at the Ballawinnie Road end of his private access lane (see photo of street sign). He tried unsuccessfully to get the name registered as an official name, but as the laneway was private, he was free to name it whatever he liked. [94]

Cut-Throat Lane Dynnyrne, Hobart SE

This steep and narrow lane used to have a clay base and it became very muddy and slippery after rain. In the 1850s there were numerous news articles referring to accidents as horses slipped on the treacherous surface and slid helplessly down the steep slope only to impale themselves on the picket fences at the foot of the hill. [95]

NB. A *Mercury* article of 1/8/1964 (page 6) states: '… *This lane took its name from a brutal murder committed there about a century ago* … ' [i.e. c1864]. This article seems to be a journalist's guess about the origin of the evocative name. The lane is now obsolete and much of it has been absorbed into the top end of Regent Street. [96]

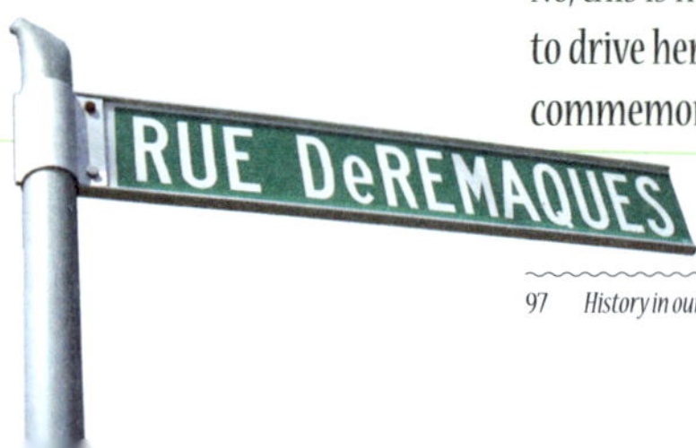

Ragged Lane Wapping, Hobart SE

This laneway once led to a Ragged School set up by a charitable enterprise to cater for pauper children in 1856 to keep them off the streets. The school sought to provide basic instruction in reading, writing and arithmetic, with a little geography, grammar and singing. They also trained the children in Christian morality, order, obedience, punctuality and thrift.

Ragged schools were commonplace in the UK in the 1800s. Ragged schools probably derived their name from the pupils as they were usually 'children dressed in rags'. Rag is a word derived from Old Norse *roggvathr* (tufted) from *rogg* (tuft of fur).

New World Avenue

Trevallyn, Launceston NT

Welcome to our brave new world! This seemingly modern name was applied around 1968 when a housing subdivision was created here by the Launceston Gas Company. The avenue was named after one of the Company's products the 'New World' gas stove. [97]

Childs Drive Old Beach, Brighton SE

No, this is not a billy-cart raceway and in fact we don't encourage children to drive here at all. The name was recommended by the subdivider to commemorate the long-time resident Childs family.

94 Advice Warren Boyles 2/4/2007, Nomenclature Board records.
95 Advised Colin Dennison March 2007
96 AOT Correspondence – 'Cut-throat Lane'

97 *History in our Streets* (Morris) p175

Queer Street west of Mangana, Break o' Day NE

It is not a street at all, but you certainly don't want to finish up here. It is a scrubby plain in the middle of nowhere where it is easy to get lost.[98]

Gay Street Oatlands SM

With gay abandon, I happily admit (to the merriment of my detractors) that I haven't got a clue how this street derived its name. The major streets of Oatlands were named after British VIPs and war heroes but nobody by the name of Gay fits this description. In addition, the name Gay is not a prominent Midland name so I am flummoxed. The street is located between the pub and the cemetery so it would experience all the extremes between gaiety and gloom. If I had a choice, I would live on the pub side of the street.

98 *Place Names of Tasmania* (Taylor & Smith)

Freaky & Leaky

'Freak Inn' Dover, Port Esperance SE

The naming of this inn was no freak, and hopefully, those who imbibed there did not freak out. It was named in 1854 after a steam ship called the *Freak* which operated out of Dover. The inn ran under the name of the 'Freak Inn' until it was sold in 1859.[99]

Drip Beach Lymington, East Huon SE

Well I never! We even have a beach for drips. This strange name was derived from an elevated spring near the shore here that once 'dripped' water over the rocks. It seems to have dried up in recent years. Interestingly, the ebbing tide here makes gurgling noises amid the rocks as if someone was pouring liquid from a bottle. Therefore, this is a good spot to have a discreet gargle.[100]

Ouse (pronounced ooze) NW of Hamilton St

This town was **not** named when early surveyors became bogged in the mud here (as one popular myth insists). Ouse Township gets its name from the River Ouse on which it is sited. The river was named c1820 after one of three rivers of this name in England.[101]

'Ouse' is a pre-Celtic river name that simply means 'water'. Therefore, Ouse River means 'water river'. What else did they expect to be in a river? Some local residents try to kid tourists that the town name is pronounced 'owse' (as in house).

Ooze Lake La Perouse, Precipitous Bluff SC

Whose lake is Ooze Lake? Who cares? The lake gets its unusual name from the globigerina ooze on its bottom that is caused by decayed protozoa (colonies of one-celled animals).[102] Any questions? This mumbo-jumbo sounds like something out of a consultant's report, so write it down in your 'useless information book'. One is tempted to add: *'You can't brew booze from the Ooze becooze, the Ooze oozes gluey ooze'*. This may be completely untrue, but we can't let a few bad puns go by, can we?

99 Dorothy Baker
100 Personal knowledge
101 Nomenclature Board records
102 *Tasmanian Place Names* (Taylor-Smith)

Lake Leake via Campbell Town, Northern Midlands

A shack at Lake Leake.
Not sure what the name means on this one!
Photo Dean Smith

A leaky lake – that's just what we need in these days of global warming. However, the name is misleading, as the lake was named in the 1880s in honour of the then recently deceased Charles H Leake, Chairman of Campbell Town Water Trust who supervised the construction of a dam here in 1884 for Campbell Town's water supply. When completed, the lake was the largest created reservoir in Australia. The lake site was previously known as Kearneys Bogs. I wonder if residents of Campbell Town realise that their drinking water comes out of a bog.[103]

Lake Leake *Photo Dean Smith*

103 Information from sign at Kalangadoo Store and from *Lake Leake* booklet at Campbell Town Information Centre

Repulsive Rivers

Repulse River

Hamilton, Upper Derwent ST

This 'repulsive' name was presumably applied by early explorers who were held up (i.e. repulsed) by this river (probably when in flood).

Dasher River

tributary of the Mersey River NW

This sounds like the name of one of Santa's reindeers, and the river acquired its name in a similar way. The reindeer was obviously a 'dasher' as it sped across the sky, so too is this river which races along the ground. The name was applied in 1826 by surveyor Henry Hellyer.[104]

Alarm River

tributary of Detention River NW

'Be alert – but not alarmed' when you try to cross this deep, fast-flowing river at one of its gorges. The river was so named by VDL Company surveyor Clement Lorymer in January 1827 when his party were afraid to cross it due to the forbidding river banks. Lorymer was unfortunately drowned three months later when attempting to cross the flooded Duck River near the present site of Smithton.[105]

Fury River Cradle Mountain CH

Named in 1828 by surveyor Henry Hellyer, after seeing the water roaring through the fearsome ravine of Fury Gorge.[106]

Linger & Die Creek

Trial Harbour, near Zeehan WC

This fantastic appellation is arguably my favourite place name. It conjures up wonderful images of prospectors who did not linger longer here than necessary particularly in winter months as it is located on an exposed plateau. Long term 'ere = hyperthermia.

Roaring Pompey Creek

Snowy Range, Upper Huon SW

Named by the well known Denny King (Melaleuca, Port Davey tin miner and a Judbury farmer) after the nickname of a long-time local resident who was renowned as a 'wild man'. 'Pompey' possibly acquired his nickname from Gnaeus Pompeius Magnus – 'The Great' Pompey' (106-48 BC), Roman General, Statesman and a member of the first triumvirate.[107]

104 *Explorers of Western Tasmania* (Binks) p42
105 *Explorers of Western Tasmania* (Binks) p59, 76
106 *Explorers of Western Tasmania* (Binks) p43
107 Nomenclature Board records and Wikipedia

Savage River Waratah WC

There is nothing savage about this river except for the intimidating name. The stream was named after Job Savage who assisted surveyor Gordon Burgess to cut a Government track from Knole Plain south of Waratah to the headwaters of the Arthur River in 1864.[108]

108 *Explorers of Western Tasmania* (Binks) p194

Sassy Creeks

Sassy Creek north Low Rocky Point WC

Sassy is defined by Webster's as 'precocious, impertinent, smart-alecky, naughty or disrespectful' but I am sure this creek does not fit into any of these categories (even if I do). In fact, the name appears to be an abbreviation for sassafras (*Atherosperna moschatum*), a slender rainforest tree that is part of the laurel family. It seems certain that this creek was lined with sassafras trees.

Tinkle Creek Portland NE

How sassy can you get? The dictionary definition of tinkle is 'a light clear ringing sound' or 'to urinate'. If the origin of the name is the latter, I pose the $64 question: who was the thoughtless (or desperate) tinkler? The answer is unknown but I hope that this creek does not flow into the local water supply.

Big Hop Creek Fingal EC

Jumping jumbucks, what have we got here? A creek that hops! I hope it doesn't jump out of its bed. In any event it is only a hop, step and a jump to Little Hop Creek located nearby. Presumably you can jump across either creek, so hop to it. The Big Hop Creek requires more effort than Little Hop.

Pistol Creek Tayatea, Arthur River WC

Trespassers don't get prosecuted here – they just get shot (or at least pistol-whipped). Nay, I jest. This is a clever name, as the stream was previously called Gunn River. Popgun Creek nearby is an associated name. Isn't it great that some place namers have a good sense of humour?

Kia Ora Creek Overland Track CH

Kia Ora Creek
Photo Brian Draper

This stream was named after nearby 'Kia Ora' Hut. Kia Ora is a Maori greeting meaning 'good health' and is used both as 'hello' and 'goodbye'.

NB. The hut reputedly gained its name during a period when the two Hartnett brothers (who were local guides) kept a New Zealand mistress at the hut for a protracted period. For those of you who are not familiar with the term 'mistress', it has been defined as: 'something between a mister and a mattress'. The brothers were no doubt practising with their visitor the awkward Maori art of rubbing noses together and generally promoting 'good health'. You may possibly wonder why the bird simply didn't take off – just remember that a Kiwi is a flightless bird. [109]

109 *Walk in the Wilderness* (Haygarth) p90

Pegs Hollow Creek St Pauls River NE

Are the pegs hollow or is it the creek? Seriously, the origin of the name is unknown, but Pegs Hollow could have been a temporary base for the itinerant tinker Thirza 'Peg' Brown. She was also known as Mrs or Mother 'One-Eye' or 'Bung-Eye Brown, the fearsome one-eyed, gypsy-like woman who used to travel around Tasmania in a horse drawn open spring cart.

She used to sell wooden dolly pegs and kitchenware made from kerosene tins in the 1920s-40s. Residents were terrified of this whip-cracking, foul-mouthed, intimidating, solid woman who was accompanied by a pack of mangy dogs. Consequently, residents would buy pegs or other supplies from her, rather than risk arousing her ire. She was also renowned for stealing boots, meat, firewood, chaff or anything that was not tied down. Her dogs used to sleep on a sack strung beneath her cart (which also may have concealed some of her stolen property). Mother Browns Bottom (Bothwell) (see page 81) may also be named after Thirza Brown.[110]

110 *Place Names of Tasmania* (Taylor & Smith), feedback from Chris Wisbey's programs on ABC Radio 12-13th May 2007

Titillating Names

The Nipple near Whitemark, Flinders Island Photo Wayne Smith

'Booby' Boobyalla NE

Boobyalla gets its name as it is sited on the Boobyalla River. 'This name was applied to the wrong river as it was an Aboriginal name for the Great Musselroe River on the eastern side of the north eastern tip of Tasmania, but it was later transferred to the present Boobyalla River further to the west. The name is descriptive of a large flowing river but the same word is also used for the coastal wattle (*Acacia sophorae*).[111] The local nickname for the township is 'Booby'.

Boob Gutter Circular Head NW

All you deviant boob watchers should get your minds out of the gutter. This name is derived from two nearby conical hills called The Boobs (for obvious reasons). The 'Gutter' is not the cleavage as the name might suggest, it is a nearby 'narrow channel worn by running water'.[112] There are two conical hills at Lymington, Port Cygnet that also share the name The Boobs.

Boobs Flat Circular Head NW

How deflating, travelling all this way to see flat boobs. However, once again the name is misleading. In fact, The Boobs are two nicely rounded conical hills (see Boob Gutter above) and the 'flat' refers to a nearby plain. Boobs Plain is an associated name and has nothing to do with 'plain' or 'ordinary' boobs.

Bubs Hill north west of the Raglan Range, Franklin River SW

It is likely that this hill was named by prospectors after its supposed resemblance to a woman's breast.

The Paps Killiecrankie, Flinders Is. BS

This is yet another peak said to resemble a substantial bosom with a prominent pointed nipple.[113] Tasmanian naming authorities seem to have had a breast fetish.

The Nipples Lake Sorell CH

There is nothing obscure about this titillating name. It is named after a pair of pointed, rocky outcrops.

'Titreet' Estate Little Swanport EC

Surely someone made a booby here? Even the Internet has never heard of a place of this name. 'Titreet' is the name of a property that was purchased by Henry Pyke in 1864. The origin of the strange name is unknown, and I refuse to speculate in case I start tittering.

111 *Tasmanian Place Names* (Taylor & Smith)
112 *Glossary of Geology* (American Geological Inst.)
113 Personal knowledge.

Milkshakes south east of Trowatta NW

This district name was derived from Milkshake Hills nearby. The hills were reputedly named by Andrew Skuja, Chief Commissioner for Forests, as he believed that they were shaped like a pair of woman's breasts.[114] However, other people believed that the hills were shaped like a sow on her back with her dugs upwards. Pigs might fly – if I was a woman, I would not like my 'milkshakes' to be compared to a sow's dugs.

Milkshake Hills   Photo Peter Daalder

114 Nomenclature Board records, ABC radio 'Nomenclature Talkback' advice

Lovely Bottoms

When referring to place names, a bottom is an archaic and amusing word for:

- 'The lowest point of land' such as a low lying valley or hollow, or
- 'The furthest position away in a downhill direction' (i.e. 'bottom paddock').

Bottoms up – let's celebrate – because Tasmania has a wonderful array of lovely bottoms. We have a Lovely Bottom, located at Carlton, and bottoms of every conceivable size and shape including Broad Bottom (Tyenna), Round Bottom (Shannon), Long Bottom (Pipers River), Deep Bottom (Ouse) etc. Feel free to look up our other lovely bottoms on the opposite page.

Bottom Fancy Reef
The Gardens, Portland NE

Are you a bottom fancier? If you are, don't waste your time coming here to satisfy your quirky fetish as this bottom is a sunken reef covered by water. But this poses the question; if the bottom is hidden, how can we be sure that it is 'fancy'? This particular bottom is said to be named after Bill Lindsay's yacht *Fancy* which beached after striking Fancy Reef (named after her). [115]

Mother Browns Bottom
Bothwell District St

'Knees up Mother Brown' was a celebrated song during WW1. Forget her knees, because in this case, we are going to focus on Mother Browns Bottom! Although the origin of the name is uncertain, it probably commemorates a favourite camping spot for the light-fingered, one-eyed, broad-bottomed, foul-mouthed gypsy-like Mother Thirza Brown. Mrs Brown was variously known as 'Peg', 'Bung-eye' or 'One-Eye'. (see Pegs Hollow Creek page 77).

Boomers Bottom
Northern Edge, Great Western Tiers Nt

Does your bottom go 'Boom!!!'? If so, we have 17 places in Tasmania named after people like you (i.e 'boomers'). Perhaps this area should have been named 'Thunder Bottom'. Please note that Baby Boomers 'boom' with less resonance than other boomers. But wait! If we get our minds above our bottoms we will find that 'Boomer' is a nickname pioneers used for kangaroos because of the 'booming' noise their tails make when they are on the move.

Officers Bottom
Clyde River, Bothwell St

This bottom is not battle scarred, not does it have stripes or pips. It commemorates Sir Robert Officer (1800-1879) New Norfolk surgeon and settler who received a grant of land on the Clyde River in 1826. [116]

Scotties Bottom Mangana, Fingal ec

The song *Donald, where's your troosers?* is probably popular here as this is an exposed bottom open to the vagaries of the weather. I have no idea who Scottie was, as there were several Scotch families living in the area? Whoever he was, Scottie would be well advised to wear his troosers here, as it can be bitterly cold in the winter.

Thistley Bottom
Township, W Beaconsfield, W Tamar Nt

Sitting bottoms on thistles is not recommended, so one shouldn't rest here. It seems obvious that this bottom was infested with Scotch thistles when named. This bottom is now a mining ghost town which served the Ilfracombe Iron Company from c1872. [117]

Hop Pole Bottom Fingal ec

This bottom hasn't been spanked with a hop pole. In any event, readers are not advised to put their bottom on hop poles, here or anywhere else. This name describes a grove of slim, tall acacias which once grew in this valley (i.e. bottom) which would have made excellent hop poles. Hop poles were used to prop up wires that supported strings for the young hop vines to climb.

115 *Tasmanian Place Names* (Taylor & Smith)

116 *Tasmanian Place Names* (Taylor & Smith)

117 *Tasmanian Place Names* (Taylor & Smith)

Ugly Bottoms

Badger Bottom Franklin River WC

This area was frequented by 'fat-arsed wombats' (at least that is what TV presenters 'Roy and HG' unkindly called wombats during the 2000 Sydney Olympics). Wombats were often called 'badgers' by early British pioneers and this is equally insulting.

Barbers Bottom Lilydale Road NT

We all know that barbers cut hair – right? So maybe this bottom was infested with hares. Did I hear a groan? The low-lying land here was owned by a farmer named Barber in the mid 1800s.

Butchers Bottom Bothwell ST

Any butcher will tell you that there is nothing like a good meaty piece of rump. However, I haven't got to the bottom of this name, so my head is on the chopping block.

Grassy Bottom Orford EC

This bottom was probably chewed by cows and is obviously a descriptive name. There is another Grassy Bottom at St Marys.

Gum Bottom Sleeping Beauty Range SE

This sounds like a sticky situation. Just when we thought it was safe to sit down, we get gum on bum. Not so, I jest. This is a descriptive name for a hollow or basin containing gum trees.

Pony Bottom Creek Sugarloaf EC

We have heard that some people are horse-faced, but I would hate to be accused of having a pony bottom. This quaint name arose when an escaped convict (possibly Thomas Prosser c1808) was captured here with a stolen pony.[118] This sounds like a shaggy pony tail to me.

Prickly Bottom Clarke Is. BS

Sitting in the water is not recommended here, because the sandstone seabed is 'prickly' under foot (or posterior).

Stumpys Bottom Ross NM

Who wants a 'stumpy' bottom? This one was named after Arthur 'Stumps' Jordan, a local snarer, who gained his nickname as a result of his endurance as a fighter when a schoolboy.[119]

Whistlers Bottom Deloraine NT

A whistling bottom!!! That's all we need. I wonder if it makes bird calls or plays 'pop' songs? Well, whether it does or not is unknown, but this low-lying area was seemingly named after one (or all) of the four Irish 'whistling' bushrangers Rogers, Reilly, Lynch and Reynolds who based themselves in the Deloraine region for a short period.[120]

The gang terrorised residents of northern and north-western Tasmania in a rampage lasting 10 weeks in 1848. One of them, John Reilly later settled near Deloraine for a few years. If the bottom was not named after Reilly or his gang it may have been frequented by chirping birds (i.e. a whistler is a species of bird). Alternatively, it may have been named after a person with the surname or nickname of 'Whistler'. The name was first recorded in 1860.

118 *Tasmanian Place Names* (Taylor & Smith)
119 *Tasmanian Place Names* (Taylor & Smith)
120 *The Whistling Irish Bushrangers Tasmania & South Australia 1848-63* (R & M Davis) p4-7, 10, 34; Nomenclature Board records

Emu Bottom Green Valley, Bagdad St

Aren't we lucky that an emu is a flightless bird? We certainly wouldn't want birds with huge bottoms flying overhead, would we? Coming back down to earth, this name preserves the fact that emus once frequented this valley. The unique Tasmanian emu was hunted to extinction by early settlers within three decades.

All W**ashed Up

Dead Mans Gulch Clarke Is. BS

This sounds like the title of an American western film. The gulch received its name when a body was washed ashore here.[121]

Golddiggers Beach Port Davey SW

This beach is not named after an actress looking for a sugar daddy. It commemorates some obscure mining activity in the late 1800s. The body of Captain Bangston of the wrecked barque *Afhild* was washed ashore here in October 1907.[122]

Guncarriage Is. now Vansittart Is. BS

The island received its strange appellation due to a miscarriage (or should that be mishap?). A ship's wooden gun carriage was found here, probably washed up from the wreck of the *Farsund* or the *Vansittart* which were both wrecked nearby.[123]

Paper Beach Lower Tamar NT

Now we don't usually have paper beaches but this is an exception. It was named by Ernest Law, a solicitor, after finding a ship's portmanteau containing lading papers washed up on the beach.[124]

Petticoat Ridge Cape Barron Is. BS

You may think that this ridge had something to do with petticoat government, but you would be wrong. A petticoat was found washed up on the shore here (probably from a wreck, unless some daring maiden here went skinny-dipping after slipping off her slip).[125]

Piano Point Cape Barren Is. BS

This place strikes a chord and sounds like a good place for honky tonks to live. The point is not ivory in colour; it gets its name from a piano washed ashore here from the wreck of the *Cambridgeshire*. It probably needed tuning afterwards.[126]

Tinderbox South of Kingston SE

Strike me lucky! This matchless name was applied after pioneer settler Joshua Fergusson found a tinderbox washed up on the beach here.[127]

Trousers Bay Flinders Is. BS

More than one person has lost their pants here. The bay was so named when a box of moleskin trousers was washed ashore from the wreck of the *Cambridgeshire* in 1875.[128]

121 *Tasmanian Place Names* (Taylor & Smith)
122 *Wrecks in Tasmanian Waters* (H. O'May) page 141, Nomenclature Board records
123 *Tasmanian Place Names* (Taylor & Smith)
124 Nomenclature Board records
125 *Tasmanian Place Names* (Taylor & Smith)
126 *Tasmanian Place Names* (Taylor & Smith)
127 Nomenclature Board records
128 Nomenclature Board records

Beerbarrel Beach St. Helens Point NE

This sounds like a good spot for a New Year's party. The beach got its strange name when beer barrels were washed ashore here from a wreck.

Well this book is all washed up too, so this is a perfect spot to conclude.
Where better, than to be sitting on a beach beside a barrel, on Tasmania's sunny north-eastern corner?

Index

About this Author

Wayne Smith is recognised as the foremost authority on Tasmania's place names and their origins. He is best known as a regular Saturday morning guest as a toponomist (see definition below*) on ABC Radio-talkback, fielding listeners' questions about nomenclature (place names). Over the past 17 years Wayne has compiled a comprehensive database of Tasmanian nomenclature and from this, he has selected this collection of quirky, unique, irreverent, unusual, outrageous and amusing place names. Where it is known, Wayne has also included the etymology (i.e. origin or meaning) of place names as this is sometimes more interesting than the name itself.

Wayne is the co-author (with Richie Woolley) of *A History of the Huon and the Far South* (published October 2004). Wayne is also co-author with the late John Taylor of an unpublished book written 1993 titled *Tasmanian Place Names*. In addition Wayne writes, edits and publishes monthly newsletters for a historical society. Unlike many historical authors, Wayne has no university education, but is an ex auditor and a qualified accountant. His training has therefore been invaluable in undertaking historical research into Tasmanian nomenclature where attention to detail is critical.

***Toponomist**: 'a historical researcher with a special interest in place names'. The word derives from *toponym* (i.e. 'place names of a region, especially those derived from topographical features'). In turn, toponym derives from the 'Greek *topos* meaning 'place' plus 'nym' meaning 'name'.

WAYNE ON HIS 'HOBBY HORSE'
FINDING THAT ELUSIVE PLACE NAME